Agricultural Risk Management Tools

Resource for the e-learning curriculum course on "Agricultural Risk Assessment and Management for Food Security in Developing Countries[1]"

Prepared By

Kisan Gunjal
Ph. D. Agricultural Economics
International Consultant

United Book Prints

Reprinted in India by:

UNITED BOOK PRINTS
(Imprint of Scientific Publishers, INDIA)
4806/24, Ansari Road, Daryaganj,
New Delhi 110 002, INDIA
E-mail: info@scientificpub.com
Website : www.scientificpub.com

Print : 2026

ISBN: 978-93-90199-89-1

Printed in India

Acknowledgements

The author would like to thank Jesús Antón, Senior Programme Manager, Platform for Agricultural Risk Management (PARM), International Fund for Agricultural Development for clear guidance and supervision. Internal review comments, questions and suggestions by him throughout the duration of this work are highly appreciated. Significant contributions of the following three reviewers, in terms of their valuable and constructive comments, are gratefully acknowledged as well.

1. Mulat Demeke
2. Vikas Choudhary
3. Gaëlle Perrin

Finally, the author would like to acknowledge the assistance of Shefaly Gunjal, Student at the University of Toronto, in conducting literature search and the preparation of the document. Errors and omissions do, of course, remain the responsibility of the author.

Acronyms

ABDI	Asian Development Bank Institute
ACE	Agricultural Commodities Exchange for Africa
ADB	African Development Bank
ADMARC	Agricultural Development and Marketing Corporation
AFD	Agence Française de Développement
AHCX	Auction Holdings Commodity Exchange
AHL	Auction Holdings Ltd.
AIC	Agriculture Insurance Company of India
AMCO	Agricultural Marketing Cooperatives
AMSDP	Agricultural Marketing Systems Development Programme
ARM	Agricultural Risk Management
BVO	Bid Volume Only
CA	Conservation agriculture
CBOT	Chicago Board of Trade
CCRIF	Caribbean Catastrophe Risk Insurance Facility
CFI	Centre for Financial Inclusion
CM	Collateral Manager
COMESA	Common Market for Eastern and Southern Africa
COSCA	Collaborative Study of Cassava in Africa
CPIS	Coconut Palm Insurance Scheme, India
CSA	Climate Smart Agriculture
CSU	Corporación de Supermercados Unidos (United Supermarkets Corporation)
CTA	Technical Centre for Agriculture and Rural Cooperation ACP-EU
DRM	Disaster Risk Management
DRR	Disaster Risk Reduction
ECX	Ethiopian Commodities Exchange
EGP	Egyptian Pound
EU	European Union
EV	Expected income and Variance frontier
eWRS	electronic Warehouse Receipt System
FAO	Food and Agriculture Organization of the United Nations
FARMAF	Farm Risk Management for Africa
FARMD	World Bank's Forum for Agricultural Risk Management in Development
FEWS NET	Famine Early Warning System Network
FOB	Free On Board
GATT	General Agreement of Tariffs and Trade
GCV	Greniers Communautaires Villageois
GFDRR	Global Facility for Disaster Reduction and Recovery
GHG	Greenhouse Gases
GIEWS	Global Information and Early Warning System on Food and Agriculture
GIZ	German *Gesellschaft für Internationale Zusammenarbeit*
GLSS	Ghana Living Standard Surveys
GMO	Genetically Modified Organism
GPRTU	Ghana Private Road Transport Union
ICN2	Second International Conference on Nutrition
ICRISAT	International Crops Research Institute for the Semi-Arid Tropics
IFAD	International Fund for Agricultural Development
IFPRI	International Food Policy Research Institute
ILO	International Labour Organization
INGO	International Non-Government Organisation
IPCC	Intergovernmental Panel on Climate Change
ITTU	Intermediate Technology Transfer Unit
LCS	Lima Credit Scheme
MACE	Malawi Agricultural Commodity Exchange

MCA	Moshi Coffee Auction
MFI	Microfinance Institution
MNAIS	Modified National Agricultural Insurance Scheme, India
MPCI	Multi-Peril Crop Insurance
MSP	Minimum Support Price
MSU	Michigan State University
MT	Metric Ton (tonne)
NASFAM	National Association of Small Farmers
NCIP	National Crop Insurance Programme
NGO	Non-Government Organisation
NRI	National Research Institute, UK
OECD	Organisation for Economic Co-operation and Development
PARM	IFAD's Platform for Agricultural Risk Management
PDS	Public Distribution System
PNAS	Proceedings of the National Academy of Sciences
PSNP	Productive Safety Nets Programs
RCS	Rural Cooperative Societies
RFE	Rural Finance Entities
RMA	USDA's Risk Management Agency
RNFE	Rural Non-Farm Economy
ROSCA	Rotating Savings and Credit Association
SACCO	Savings and Credit Co-operative
SADC	Southern African Development Community
SAFEX	South African Futures Exchange
SBI	State Bank of India
SIDO	Small Industries Development Organization
SOFA	State of Food and Agriculture of FAO
SRM	Social Risk Management
SSA	Sub-Saharan Africa
SME	Small and Medium-sized Enterprise
SWC	Soil and Water Conservation
TCG	Trade and Commodity Group
TWLB	Tanzania Warehouse Licensing Board
UNCTAD	United Nations Conference on Trade and Development
UNIDROIT	International Institute for the Unification of Private Law
UNISDR	United Nations Office for Disaster Risk Reduction (renamed from United Nations International Strategy for Disaster Reduction)
USAID	United States Agency for International Development
USD	United States Dollar
USDA	United States Department of Agriculture
WB	World Bank
WBCIS	Weather Based Crop Insurance Scheme
WED	Women Entrepreneurship Development
WFP	World Food Programme
WHO	World Health Organization
WII	Weather Index Insurance
WRS	Warehouse Receipt(s) System
WYX	Weather Yield Index
ZAMACE	Zambia Agricultural Commodities Exchange
ZNFU	Zambia National Farmers' Union
ZSIC	Zambia Insurance State Company

Table of Contents

Case Studies

Overview of the Course

This learning module is part of a broader joint effort of international agencies, experts and practitioners to develop a curriculum on "Agricultural Risk Assessment and Management for Food Security in Developing Countries". A curriculum outline was developed after a broad consultation led by IFAD and FAO in Rome in July 2015, and the Platform for Agricultural Risk Management (PARM/IFAD) has supported the development of this module that corresponds with module 3 of the curriculum[1].

Objectives:

The main objective of this course module is to familiarize the learner with a vast array of tools potentially applicable for agricultural risk management by providing the basic knowledge of each with respect to their content, pros and cons, and suitability in given specific risk situations.

Target Audience:

Based on the type of functions performed and the level of involvement at the micro, meso and macro levels of the principal stakeholders, a potential target audience for this informative technical e-learning course is mostly policy makers dealing with risk program development and implementation. There is a need for government participation as they are likely to play an important role in a broader strategy involved in tools such as CSA, non-farm diversification, assets based strategies, contract farming, food grain reserves, disaster assistance programs, social protection, safety nets, etc. However, considering a broader perspective, the potential target audience can be divided into the following three main groups:

1. **Policy Group** including policy makers and planners and other government officials (mainly macro-level stakeholders, functioning at the overall sector level).
2. **Farm Group** including farmers, farm management advisors, farmer organizations and community organizations (mainly micro-level stakeholders, functioning at the farm/firm/household level), and
3. **Services Group** including service and input providers and other intermediaries, particularly those dealing with agricultural risk management products and services (mainly meso-level stakeholders).

In addition, people from academia, research and other knowledge seekers of the subject may be interested in this course. Finally, it should be kept in mind that not all tools will be relevant to all groups.

Context and Scope of the Course:

Risks in agriculture arise from a variety of sources. Therefore, to deal with them effectively, there are various types of tools available that are suitable to deal with either a single specific risk or multiple risk situations. Some risks are systemic (e.g. droughts and floods), affecting a large part of a geographic area while some are non-systematic or idiosyncratic. Some risks can be managed with traditional farm management solutions, some risks are insurable through markets and some can be minimised with government support. Therefore, when selecting the appropriate tool(s) it is

---

[1] For further details see FAO/IFAD report, Curriculum Outline E-learning course on: "Agricultural Risk Assessment and Management for Food Security in Developing Countries" by Kisan Gunjal, 7 September 2015.

necessary to consider all possible scenarios involving farming risks, their correlation with each other, farmers' strategies and government policies with a holistic approach[2].

The entire eLearning course developed by FAO/IFAD deals with risk assessment and management and includes the below mentioned four modules. This particular document covers only Module 3, dealing with various instruments useful in managing risks found in agriculture.

- Module 1: Understanding the Risk Environment in Agriculture
- Module 2: Assessing Risk in Agriculture
- Module 3: Agricultural Risk Management Tools
- Module 4: Agricultural Risk Assessment and Management Strategy, Policy And Mainstreaming

This course, that provides the basic knowledge of over a dozen tools, is not a "how-to" practical guide on the application of the various risk management tools. This is because they are highly complex and would require the intervention of experts in each individual field. Rather, the scope of this course is limited to creating a basic understanding of various management tools available in different risk situations. Furthermore, it should also be emphasized that the application of risk management tools is an integral part of a comprehensive risk management strategy which generally consists of the following iterative sequential process proposed by the World Bank[3].

- Risk Monitoring
- Risk Assessment/Reevaluation
- Solutions Assessment
- Operationalizing & Mainstreaming Risk management
- Implementation

Decision-makers as the principal target audience, need to know about these tools and how they should fit together in a comprehensive risk management package. This last point is the aim of agricultural risk assessment that prioritizes risks and proposes risk management packages to manage risks of high priority.

In general, risk management tools can be grouped into three categories based on the approach taken to deal with them, i.e. prevention (or risk reduction), mitigation and adaptation (or coping). Prevention strategies aim to eliminate or reduce the occurrence of disaster-causing hazards, mitigation strategies aim to reduce and manage the potential adverse impact of a disaster before, during and after its occurrence, and adaptation strategies deal with measures of coping with and relieving post-disaster impacts.

Although catastrophes and disasters play a significant role in creating risk in agriculture, the strategies of risk management are not necessarily limited to disaster risk management. For example, climate smart agriculture activities, commercial weather index insurance policies or the price hedging tool are used as ARM instruments with or without underlying disaster scenarios. Good risk management strategies will reduce the number of risks that become disasters.

The various tools presented in this course cross-over these non-distinct boundaries between the three categories mentioned above, some even spanning over all three. Hence, the ARM tools

[2] For further details of the concept see OECD, 2009. Managing risk in agriculture: a holistic approach.

[3] World Bank, Jan. 2016. Agricultural Sector Risk Assessment: Methodological Guidance for Practitioners Agriculture Global Practice Discussion Paper 10.

covered in this module are categorized on the basis of their operational modalities, into the following four domains or lessons:

- Lesson 1 - On-Farm and Community Level Risk Management Tools,
- Lesson 2 - Finance Related Risk Management Tools,
- Lesson 3 - Market Related Risk Management Tools, and
- Lesson 4 - Government-based Agricultural Risk Management Tools

The scope of this course is the range of key tools available in various farming risk situations. However, it is not intended to cover all possible strategies and instruments. This course is mostly intended to be the first step towards being an expert practitioner in the field as the depth of coverage is limited to providing only an appreciation and basic understanding of the tools discussed.

The topics covered in this course focus on what can be done at farm, community, regional and national levels to manage risks in agriculture. Geographically, the primary focus of this course is on Africa, although pertinent examples from other developing regions such as Asia and Latin America are used.

Each of these major categories of ARM tools are discussed in the corresponding four lessons. Each lesson is further divided into three specific tools, covering 12 relevant tools of ARM in total. It is recognized that this may not be an exhaustive list of all possible ARM tools, but it is hoped that together they cover major instruments, programs, policies and/or practices that are useful in reducing, sharing, transferring or mitigating risk in agriculture. The learner can pick and choose the most relevant ones to suit his/her specific situation or interest.

Each lesson in this module covers a brief chapeau and a list of key words followed by the details of the three specific tools included in that category. The description of each tool follows the following seven point format:

1. What is it? – Concepts and Characteristics
2. Suitability of the tool (for whom, for what type of risk, for which crops or livestock, where and under what conditions)
3. Advantages (general positive aspects - pros)
4. Disadvantages (general negative points - cons)
5. Examples and/or Case studies
6. Assessment test for students, and
7. List of relevant references

LESSON 1
On-Farm and Community Level Risk Management Tools

Learning objectives:

At the end of the lesson, learners should be able to understand the nature of various on-farm tools relevant for farmers or communities, judge their suitability for a given area or farm situation, assess their advantages/opportunities and disadvantages/constraints, and appreciate complementarities of the selected tool with other tools.

Farmers around the world cope with inherent risks in agriculture by following traditional as well as more modern on-farm or off-farm practices. Although tools outlined in this lesson require management responses on the part of the farmer, they may also need involvement and interventions by the community and government policy makers. Farmers themselves need to be empowered by the government, community and others to take responsibility of risk management. Policy actions should exploit the correlations between farming risks and provide optimal solutions.

This lesson pulls all major farm management practices together that can be considered as tools for risk management. These are as follows:

Tool 1: Climate Smart Agriculture
Tool 2: Crop and Enterprise Diversification, and
Tool 3: Asset and Income Based Strategies

Climate change is perhaps the most pressing issue of our time, demanding appropriate actions from all of us, but most prominently from those managing our agricultural production systems. Fortunately, as we collectively understand the impact of the changing climate better while also gathering an increasing amount of information on how to mitigate and adapt to this change, there is a body of actionable knowledge/technologies we can potentially employ.

This tool therefore presents various practices from conservation agriculture, soil and water management and improved animal husbandry into a package that is most popularly known as climate smart agriculture. Although it does not cover an exhaustive list of potential activities in this domain, it does include the major ones and gives an idea of how to mitigate and adapt to climate-led changes on one hand, and how to reduce the impact of agriculture on the climate itself on the other.

The other set of farm strategies involve the formulation of farm production plans, more specifically, crop, enterprise and income diversification based on the principle of minimizing risks to family income and livelihood. This strategy extends beyond on-farm planning. In Tool 3, it is recognized that asset and income diversification strategies available to farmers as well as opportunities provided through a strong non-farm rural economy play a very important role in managing on-farm risks. The enabling environment is generally provided by a local/regional community, other stakeholders and public decision makers to improve the livelihood of those dependent on agriculture.

A description of each tool provides the basic knowledge of the main concepts, evidence from the literature of the tool's performance and an example or case study of a practical situation where it has been employed.

Finally it should be mentioned that this lesson does not deal with all possible tools. For example, community risk management tools or practices, large-scale irrigation schemes, asset liquidity, remittances from various sources, among others, are not covered explicitly.

Key words for Lesson 1:

1. *asset allocation,*
2. *carbon sequestration,*
3. *climate smart agriculture,*
4. *conservation agriculture,*
5. *crop diversification,*
6. *crop rotations,*
7. *enterprise diversification,*
8. *farming systems,*
9. *greenhouse gases,*
10. *income diversification,*
11. *minimum tillage,*
12. *niche crops,*
13. *non-farm income,*
14. *soil conservation,*
15. *sustainability*

Tool # 1: Climate Smart Agriculture

1. What is it? – Concepts and Characteristics

The objective of this tool is to provide a basic understanding of various modern and traditional practices in the domain of "climate smart agriculture" (CSA) and to explain how they can aid in the task of managing agricultural risk.

The climate is changing globally and is increasing risks in agriculture with important regional variations. The data from the United Nations International Strategy for Disaster Reduction (UNISDR) shows a dramatic increase in the occurrence of natural disasters, particularly of hydro-meteorological events, during the last century. The steady worsening of climatic conditions such as global warming and the increasing frequency of extreme weather events, including droughts and floods, have a negative impact on crop yields and food production.

An increase in climate variability poses major challenges to agriculture and food security because of its effects on the basic elements of food production – soil, water and biodiversity. Hence, scientists, other experts and policy makers have been working to develop new and innovative measures to increase the adoption of such measures to help mitigate the negative impacts of climate change.

Climate-Smart Agriculture (CSA)[4] is defined by FAO as "an approach that helps to guide actions needed to transform and reorient agricultural systems to effectively support development and ensure food security in a changing climate. CSA aims to tackle three main objectives: sustainably increasing agricultural productivity and incomes; adapting and building resilience to climate change; and reducing and/or removing greenhouse gas emissions, where possible." (FAO 2013a, b). Similarly the World Bank (2014) considers CSA as an approach to managing landscapes (cropland, livestock, forests and fisheries) that aims to achieve the above mentioned outcomes.

The CSA practices, carried out individually or in various combination, can result in climate change adaptation and mitigation benefits, as well as higher and more stable yields and thus reducing risk. The overall negative impacts of climate change on production, incomes and well-being can be avoided with these practices in tandem with improved weather, early warning systems and disaster risk management approaches.In general, a CSA strategy that deals with both short-term inter-annual production risks and the adaptation to long-term climate change risks, bridges the short and long term efforts to achieve the resilience of agricultural systems.

In addition to the use of improved crop varieties, productivity enhancing inputs and modern practices, the following set of activities are considered to be the three main pillars of CSA:

1. **Conservation agriculture,**
    a. Minimum tillage,
    b. Permanent crop cover with crop residue mulching
    c. Crop rotations,
    d. Intercropping,

---

[4] Another related concept which may go beyond agriculture is **climate change adaptation** (CCA), defined by the United Nations Framework Convention on Climate Change (UNFCCC) as "adjustments in natural or human systems in response to actual or expected climatic stimuli or their effects that moderate harm and exploit beneficial opportunities. This can include: (a) adapting development to gradual changes in average temperature, sea level and precipitation; and (b) reducing and managing the risks associated with more frequent, severe and unpredictable extreme weather events" (UNISDR, 2010).

   e. Planting pits, etc.

2. **Soil and water conservation**
   a. Use of farm yard manure,
   b. Terraces and bunds,
   c. Rain water harvesting structures and systems,
   d. Use of water efficient irrigation systems and crops, etc.
3. **Improved livestock management practices, etc.**
   a. Improved feed management,
   b. Optimal livestock manure management,
   c. Desired destocking of livestock,
   d. Switching to livestock species or breeds that are more adapted to water scarcity and are resistant to disease and pests,
   e. Pasture management, and
   f. Sustainable fisheries and aquaculture.

Conservation Agriculture (CA)

CA, seen as a component of CSA, encompasses three primary principles dealing with crop cultivation - 1) minimum soil disturbance, 2) minimum bare, uncovered soils (i.e. having a permanent soil cover), and 3) minimum artificial fertilizer application by use of crop rotation (FAO 2013a). The combination of these practices is aimed at reducing wind and soil erosion, increasing water retention, improving soil structure, aeration and fertility to increase crop yields sustainably and building the resilience of farm systems.

In practice, minimum soil disturbance may include a wide range of tillage from **zero tillage** to as little tillage as required depending on the local circumstances. The ultimate goal is to minimize soil disturbance. A study by Derpsch et al. (2010) showed that the area under no-tillage farming worldwide has grown from 45 million ha in 1999 to 111 million ha in 2009, with the fastest adoption rates found in South America. The reduced labour and time on tillage under CA can allow farmers to manage timely plantings to take advantage ofirregular rains.

Permanent soil cover involves planting cover crops, covering the soil with crop residue from previous plantings, or using other types of mulch on the field even after harvest. Cover crops and mulch can help manage soil temperatures, reduce evapotranspiration, incidence of weeds, and promote a healthy bio-balance in the soil, thus helping to improve crop yields.

In areas of low and erratic rainfall, a variation of conservation agriculture is practiced where maize or other crops are planted in a "**fertility pit**" filled with organic matter, such as compost or manure. The pit also allows runoff water to be captured and conserved for better plant growth. The organic cover minimises the loss of water from evaporation. Experiments in Zambia have proven beneficial in getting higher yields and sustaining a steady harvest over time.

Crop rotations and intercropping involve a farm planning process whereby crops such as cereals, pulses, roots and tubers, etc. are grown in a deliberate sequence or mix, season after season. Ideally, nitrogen-fixing crops/plants are included in the cycle to increase soil fertility, especially for subsequent crops leading to increased crop yields. An example could be a maize-cotton-cowpea rotation in Southern Africa. Monoculture, on the other hand, is the repeated planting of the same crop in the same field year after year; this has an adverse effect on soil structure and productivity.

The data and experience from Malawi has shown **agroforestry,** involving selected leguminous tree and shrub species planted during fallow or intercropped with food crops, can help maintain/improve soil cover, nutrient levels, soil organic matter, water filtration, and provide a secondary source of food, fodder, fibre and fuel (FAO, 2013).

The literature (FAO 2014a; Derpsch et al. 2010) suggests that CA offers a significant potential in Africa to improve food security and adapt to climate change by increasing soil fertility, erosion control, soil absorptive capacity, and easing drought stress due to improved water retention. An FAO study (2013) found a very strong relationship between the variation in rainfall and the adoption of CA practices in Zambia, suggesting that farmers use minimum tillage/planting basins as a strategy to mitigate the risk of rainfall variability, providing evidence of a synergy between the CSA practices and adaptation to climate variability.

Soil and Water Conservation

Long term climate data and various model simulations indicate that water crises in large parts of Africa are imminent. Therefore, proper **water management** that helps to capture more rainfall, make more water available to agriculture, and increase water use efficiency, is crucial to augment agricultural production sustainably.

The central focus of **soil and water conservation** (SWC) practices is to reduce or eliminate soil erosion and degradation. A related goal is to manage water quality through addressing rainfall runoff and factors influencing groundwater quality. Biological practices include improved fallows, cropping patterns, and manuring/mulching. Mechanical structures for water harvesting/conservation can be appropriate additions to land where biological methods are not sufficient in controlling erosion. Such structures include terraces, ditches, spillways, bunds, silt fences and surface mats, among others. Furthermore, rainwater harvesting and recycling with the use of water ponds or rooftop rainwater collection systems are highly effective and efficient especially in erratic rainfall and hilly areas (FAO 2014c).

Improved livestock management practices

In addition to crop farming, livestock activities are shown to have a significant impact on greenhouse gases (GHGs) and environmental degradation if left unchecked. There are several improved livestock management practices that are aimed at increasing food security, enhancing farmers' resilience and reducing GHGs (see details in World Bank/FAO/IFAD, 2015; Bernier et al. 2015).

Finally, agricultural soils may also act as a sink or source for CO_2 and many agricultural practices can potentially mitigate GHG emissions. In particular, CSA can lead to the slowing down of the loss of carbon sequestered in soil, as well as increased fertility. A higher biomass can increase the potential for carbon sequestration. Reducing GHG emissions at the farm level will have a measurable positive long term impact on the climate if practiced by all farmers (Smith et al. 2008).

2. Suitability of the tool

CSA practices are generally suitable for all farmers around the world. They can be suitable for both smallholder as well as largescale farmers to adapt to climate change and reduce risks to crop yields and production. Underlying CSA practices can increase water availability for irrigating crops, improve water retention and reduce soil erosion, especially on lands with a slope.

This tool is especially important where soil, water and other resources are scarce, in poor condition or under severe climatic stress, especially in developing countries. The greatest gains are possible under the more vulnerable/fragile environments. The literature (e.g. FAO 2011) also suggests that soil and water conservation structures, one of the important planks of CSA, are highly suitable and likely to produce relatively more benefits in mountainous farming areas and sub-humid regions.

The benefits of practicing no-tillage in semi-arid areas seem to be higher on severely degraded soils because of low organic matter content and poor physical soil structure (Acharya et al. 1998).

This tool is suitable where property rights and land tenure security, especially for smallholder farmers (FAO, 2013; Byamugisha, 2013) are clear and well defined.

3. Advantages

There are numerous empirical scientific studies that have demonstrated the positive impacts of various elements of CSA listed above on the level and stability of crop yields, thereby helping to reduce the yield/production risk especially in the long run (for a comprehensive review see FAO 2011). The benefits of various practices accrue, given the ability of CSA, to improve the capacity of smallholder agriculture to be more productive in the medium to long run by achieving the following intermediate gainful results:

- Reduced wind and soil erosion,
- Improved soil structure,
- Increased soil fertility,
- Better control of pests, diseases and weeds
- Higher water retention,
- Reduced labour use for land preparation
- Increased rain water use efficiency
- Mitigation of downstream flooding,
- Reduced sedimentation of waterways,
- Slowdown of the loss of carbon sequestered in soil,
- Enhanced biodiversity, etc.

4. Disadvantages

Although CSA offers significant benefits in principle, the timing of these and the resources required as investments can be negative factors constraining its adoption. Despite the potential, the adoption of such practices remains generally low, particularly in sub-Saharan Africa (SSA). As reviewed by FAO (2011), there are several limitations which may slow down or prevent the adoption of CSA, especially by smallholder farmers. These are:

- In many cases, there is a time lag of four or more years before the full benefits of CA are realized.
- The costs and investments are generally incurred up front. These investments, for example, in the case of CA (for no-till drill, fertilization equipment, herbicides for weed control, etc.) or the building of water conservation structures, can be substantial and a major deterrent for many smallholder farmers.
- Given the long-term nature of the investment and benefits accruing to these practices, security of tenure becomes an important consideration.
- For conservation agriculture programs that promote the use of crop residues, the opportunity cost of those residues is an important determinant in the adoption of CA.

- Finally, a successful implementation of CSA techniques may require adequate management and technical skills. This depends on the ability and willingness of the farmers as well as access to training, guidance and financial support.

All these limitations are important in practice but can be overcome with the right priorities, a set of appropriate policies and concerted efforts by all stakeholders.

5. Examples/Case studies

EXAMPLE: Women's Adoption of Transformational Changes in Agricultural Practices: Evidence from Bangladesh, Kenya, Senegal, and Uganda

Transformational CSA practices include those that contribute to diversified livelihoods, aim to buffer the household against climate change, increase assets, take a relatively long time for benefits to accrue, and require substantial investments of time, labor, or cash. An intrahousehold study in four countries in East and West Africa and South Asia reports that improvements in women's access to information and credit enhance the likelihood that they will adopt new, transformational CSA practices. Local agricultural groups are key sources of information on CSA, and for women, in particular, they are also important for sharing labor. The study also finds that although male farmers supported by extension officers are more likely to make transformative changes, female farmers who receive extension advice are not. The risks and trade-offs specific to women as they decide whether to invest in new practices imply that without more targeted support and services that address women's needs, the challenge of achieving the multiple goals of CSA will remain significant.

Source: Cited in FAO, IFAD, World Bank. 2015. Page 6. http://www.fao.org/3/a-az917e.pdf

CASE STUDY 1: Climate-smart villages in Lower Nyando, Kisumu, Kenya (*World Bank*)

The main stakeholders in this World Bank case study are smallholder farmers in Lower Nyando, Kisumu, Kenya, facing variable weather patterns, high GHG emissions, soil erosion, declining soil fertility, high poverty rates and an increasing human population. This case study provides an example of managing farms more sustainably at a village level with the help of other finance based risk management strategies involving the use of agroforestry, growing high quality seeds, etc. An innovative application of ICT mobile phones for CSA advice is also used to provide access to index based insurance and other financial tools. The study identifies the weaknesses/constraints such as limited exposure to innovations, poor management of pests, diseases and livestock, as well as the fact that CSA technologies are knowledge intensive. It also suggests that a link between the private sector, farmers and markets through microfinance will be beneficial.

Case study of CSA in Kenya:
Climate-smart villages in Lower Nyando, Kisumu

The Lower Nyando climate-smart village (CSV) is one of the 15 villages established in 2011 by the CGIAR Research Program on Climate Change, Agriculture and Food Security (CCAFS). A CSV is an experimental plot aimed to enhance farmer's resilience through improved agricultural productivity, income levels and mitigation of greenhouse gases from agriculture.

Smallholder farmers in Lower Nyando are confronted with variable weather patterns that have caused crop failures, undermining livelihoods and threatening local food security. At the same time, common farming practices have been a significant contributor to agricultural GHG emissions, further intensifying the impacts of climate change. Furthermore, the region is also facing problems related to soil erosion, declining soil fertility, high poverty rates, low farm labour productivity, and a growing human population.

In response to these challenges, CCAFS initiated the CSV in 2011 in an attempt to adapt agricultural activities to changing climate conditions and to ensure a stable and improving food supply with additional mitigation benefits. CSVs are community driven, participatory, and inclusive. Stakeholders test climate-smart services, such as tailored weather agro-advisories for farm planning and management, allowing for continued improvement. These services are delivered through ICT mobile telephony, enabling farmers to buy index-based insurance to protect them in the event of extreme weather.

As a result, about 22 tree nurseries have been established to supply over 50,000 high-quality tree seedlings with a 75% tree survival rate. Women own more than half of these nurseries and sell seedlings for extra income, allowing them to send their children to school and buy more nutritious food for their family. More than 11,000 households in the CSV have joined self-help groups, of which about 70 to 85% of the active members are women. Farmers have discovered the value of agroforestry, with alleys of maize, sorghum and other crops sandwiched between rows of multi-purpose trees that stabilize and enrich the soil. Peer learning, where certain farmers host neighbours on their farms for knowledge-sharing events, has promoted cross-fertilization of ideas in the region.

Even though several opportunities exist with the CSVs, CSA technologies are knowledge intensive, and farmers have limited exposure to agricultural innovations. Similarly, pest and disease management is poor, slowing the uptake of improved livestock management practices. To maximize available opportunities, links between farmers and the private sector and markets need to be created, via sustainable initiatives such as microfinance.

An agroforestry field, Kisumu, Kenya
Photography: P. Casier (CGIAR)

Source: World Bank. 2015. Climate-smart Agriculture in Kenya, Climate Smart Agriculture Profiles. Page 10. http://sdwebx.worldbank.org/climateportal/doc/agricultureProfiles/CSA%20KENYA%2NOV%2018%202015.pdf

6. Assessment Test for Tool # 1: Climate Smart Agriculture

To manage the dominant risk in each of the following typical scenarios in a given local area, which two climate smart agriculture (CSA) practices (from the given list at the bottom) would you recommend as the most appropriate ones for farmers to follow?

1. A dry farming area where the threat of frequent drought is very high.

2. A hilly area where rainfall is relatively high but irregular and causes a high risk of soil erosion, landslides, etc.

3. An area where the prolonged use of monoculture has eroded soil structure and an increasing application of fertilizer is needed to maintain yield levels.

4. In a predominantly pastoralist area, an increasing loss of vegetation and rise in soil erosion and degradation, over time, is experienced.

A. Crop rotations/mixed cropping/intercropping
B. Improved pasture/vegetation/agroforestry management
C. Mulching/green manuring
D. Rainwater harvesting and conservation
E. Reducing number of animals per hectare of land
F. Reforestation/Agroforestry
G. Conservation agriculture with fertility pits/minimum tillage
H. Terracing for crop cultivation

Answers

1. G., D.
2. H., Г.
3. A., C.
4. E., B

7. References:

1. **Acharya, C., Kapur, OC, Dixit, SP.** 1998. Moisture conservation for rainfed wheat production with alternative mulches and conservation tillage in the hills of north-west India. Soil Tillage Res. 46 (3-4): 153-163.

2. **Bernier Q, R. Meinzen-Dick, P. Kristjanson, E. Haglund, C. Kovarik, E. Bryan, C. Ringler, and S. Silvestri.** 2015. "What Influences Awareness and Adoption of Climate Smart Agricultural Practices? Evidence from Kenya." CCAFS Working Paper 79. CGIAR Research Program on Climate Change, Agriculture and Food Security (CCAFS), Copenhagen, Denmark.

3. **Byamugisha, F. F.** 2013. Securing Africa's Land for Shared Prosperity: A Program to Scale Up Reforms and Investments. Washington, DC: World Bank.

4. **Derpsch, R., Friedrich, T., Kassam, A., & Hongwen, L.** 2010. Current status of adoption of no-till farming in the world and some of its main benefits. International Journal of Agricultural and Biological Engineering, 3 (1), 1 -25.

5. **FAO.** 2015 Helping farmers cope with climate change.

6. **FAO.** 2014a. Climate smart agriculture and resource tenure in sub-Saharan Africa: a conceptual framework, by McCarthy, N., Brubaker, J.

7. **FAO.** 2014b. FAO success stories on climate-smart agriculture CSA on the ground.

8. **FAO.** 2014c. Managing climate risks and adapting to climate change in the agriculture sector in Nepal, by Selvaraju, R. Climate, Energy and Tenure Division (NRC), Rome, Italy.

9. **FAO.** 2013. Adoption and intensity of adoption of conservation farming practices in Zambia, by **Arslan, A., McCarthy, N., Lipper, L., Asfaw, S., & Cattaneo, A.** ESA Working paper No.13-01.

10. **FAO.** 2013a. Climate-Smart Agriculture Sourcebook. Rome, FAO.

11. **FAO.** 2013b, Climate-Smart Agriculture? A review of current practice of agroforestry and conservation agriculture in Malawi and Zambia by Kaczan, D., Arslan, A., & Lipper, L. ESA Working Paper No. 13-07. Rome, FAO.

12. **FAO.** 2012. *Conservation agriculture and sustainable crop intensification: a Zimbabwe case study,* Integrated Crop Management Vol.17, by Marongwe, L. S., Nyahumbo, I., Kwazira, K., Kassam, A., Friedrich, T.

13. **FAO.** 2011, Dec. Climate-Smart Agriculture: Smallholder Adoption and Implications for Climate Change Adaptation and Mitigation, by McCarthy, N., Lipper, L., Branca, G. Mitigation of Climate Change in Agriculture (MICCA) Series 4.

14. **Hobbs, P. R., Sayre, K., & Gupta, R.** 2008. The role of conservation agriculture in sustainable agriculture. Philosophical Transactions of the Royal Society B, 363, 543-555.

15. **Smith, P., Martino, D., Cai, Z., Gwary, D., Janzen, H., Kumar, P., et al.** 2008. Greenhouse gas mitigation in agriculture. Philosophical Transactions of the Royal Society B: Biological Sciences , 363 (1492), 789-813.

16. **Teshome, A., Rolker, D., & de Graaff, J.** 2013. Financial viability of soil and water conservation technologies in northwestern Ethiopian highlands. Applied Geography, 37, 139-149.

17. **Thierfelder, C., Wall, P. C.** *The Importance of Crop Rotations. A Leaflet on crop rotation and green manuring.*

18. **Thierfelder, C., Cheesman, S. & Rusinamhodzi, L.** 2013. *Benefits and challenges of crop rotations in maize-based conservation agriculture (CA) cropping systems of southern Africa,* International Journal of Agricultural Sustainability, 11:2, 108-124 2013.

19. **UNISDR.** 2010. Local Governments and Disaster Risk Reduction: Good Practices and Lessons Learned, United Nations Secretariat to the International Strategy for Disaster Reduction (ISDR), Geneva.

20. **World Bank.** 2015. Climate-smart Agriculture in Kenya, Climate Smart Agriculture Profiles, Climate Change Knowledge Portal.

21. **World Bank.** 2015a, Oct. Agricultural risk management in the face of climate change. Agriculture global practice discussion paper 09. IBRD.

22. **World Bank, FAO and IFAD.** 2015b. Gender in Climate-Smart Agriculture: Module 18 for Gender in Agriculture Sourcebook. http://www.fao.org/3/a-az917e.pdf

23. **World Bank.** 2014, Nov. Foster Climate-Smart Agriculture. Brief. http://www.worldbank.org/en/topic/agriculture/brief/foster-climate-smart-agriculture

24. **World Bank.** Forum for Agricultural Risk Management in Development (FARMD). The World Bank's Risk Mitigation for Smallholder Agricultural Production in the Caribbean Course. Module 1.2: Adaptation Measures for Weather Risks. Online training course at - https://www.agriskmanagementforum.org/content/world-banks-caribbean-smallholder-farmer-course

Tool # 2: Agricultural Diversification

1. What is it? – Concepts and Characteristics

The primary focus of this tool is on how to take advantage of modern farming systems at the farm level so that the risks associated with crop and livestock activities are minimized and the sustainability and resilience of the farm family livelihood is increased. The diversification of agricultural activities/enterprises is one of the important traditional risk management, self-insuring mechanisms adopted by farmers around the world.

Agricultural diversification is defined by ICRISAT as the "re-allocation of some of a farm's productive resources (land, capital, farm equipment, paid labour, etc.) into new activities to reduce risk against climate variability, agricultural price volatility and other factors at the same time as generating additional income." The primary objective of agricultural diversification is to spread the risk in farming while maintain the highest possible level of income for the farm family's livelihood. There are two main types of diversification strategies in agriculture:

- Crop diversification, and
- Farming enterprise diversification.

Crop diversification, through multiple crop cultivation, reduces the risk of crop production loss or its value due to adverse weather, pests and disease attacks, or unfavourable market conditions for any one crop. It may include choosing multiple crops, either by adopting crop rotation over multiple seasons or a mixed/intercropping system in the same season. In the former case, the income variability as a measure of risk is viewed over that multi-season period rather than on a strictly annual basis.

Enterprise diversification refers to a broader farm operations plan that includes several farm enterprises, not just staple or traditional crops but also other cash or niche crops (organic produce for example), livestock, aquaculture, apiculture and possibly further basic farm level processing. These can be seen as different enterprises and can contribute to increasing productivity in a synergetic/symbiotic way, potentially offsetting losses in one with gains from others. Thus, it can lead to a shift to a higher and more stable farm income.

The basic principle underlying any diversification plan, whether it is in agriculture, the industry or in maintaining a personal investment portfolio, is the inclusion of activities whose earnings (based on production and net prices) are negatively or weakly positively correlated with each other. The total farm enterprise risk (or portfolio risk) is measured by the sum of variance contributed by each activity plus their covariance with each other. (See details *Markowitz 1959;* OECD 2009). Therefore, if the covariance is negative, (meaning if the incomes of the two activities are negatively correlated) total variability (the risk) associated with the total net income would be lower. Basically, this means that the lower income from some activities would likely be offset by a higher income from some others.

The theoretical considerations behind diversification are explained in Box 1. In practice, however, as stated by Blank (1990) "one strategy for portfolio creation is to start with the highest returning crop then continue to add crops or other enterprises with returns that have the greatest amount of negative correlation with the first product and/or the portfolio." It should also be mentioned that the most common diversification strategies include crop and livestock activities based on their mutual synergetic relationships.

Box 1. Portfolio theory behind diversification (applicable for Tool 2 and 3):

Farm planning with an objective to maximize the expected net income (E) for a given level of risk/variance (V) (or conversely, minimizing risk for a given level of expected income) can produce an efficient portfolio of activities. When this is repeated at various given levels of the other factor (E or V), the set of plans produced is called an efficient frontier or the EV frontier as shown in the figure below. This EV frontier curve is generally concave to the vertical V axis, meaning as the expected income from the portfolio increases, the risk (variance) increases too but at an exponential rate. The EV frontier is generated by mathematical quadratic programming for a given farm.

In order to decide which specific plan among the set of efficient plans is the right one for a given farmer, one has to look at his/her risk preference (risk acceptance or risk tolerance). Based on the risk preferences of the farmer (risk averse or risk neutral) one optimal plan can be chosen (Markowitz 1959, Blank 1990). Typically, a risk aversion preference generates an upward sloping curve with its slope and shape depending on the degree of risk aversion of the farmers. The risk aversion is the most common behaviour of small holder farmers.

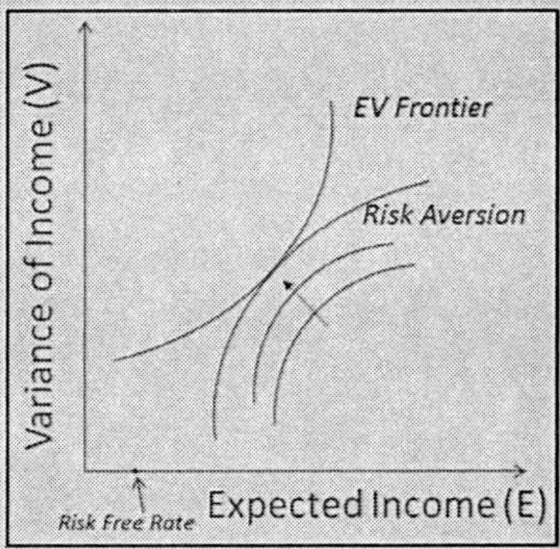

In theory, the tangency point between the risk preference curve (highest indifference or iso-utility curve) and the EV frontier curve produces the best possible outcome; the most optimal/desirable diversified farm plan. Empirically, risk preference curves are determined by experimental/empirical economics research working with different decision making farmers.

Various triggers for farm diversification include emerging market opportunities due to changing consumer tastes and preferences (high value products, organic produce, etc.), demographics (including urbanization, ethnic population mix creating a demand for niche markets overtime, etc.), and export market potential.

Finally, it should be kept in mind that diversification is a dynamic concept. As the farmer's circumstances (climatic, personal and market conditions) change, the ideal diversified plan will also change overtime to achieve a better balance of risk minimization and income maximization.

2. Suitability of the tool

This tool is suitable for all farmers, especially those that have sufficient land to cultivate different crops and/or undertake livestock activities. This tool is primarily useful to manage farm production risk.

Many farmers around the world, especially smallholder farmers, integrate crops and livestock to reduce risk and to improve their efficiency in resource use and sustainability of the natural resource base. Smallholder farmers in harsh environments, like the semi-arid tropics, use diversification to better manage the risks for larger income opportunities and better resilience of the farming system in the long-term.

A study (Mesfin 2011) from the eastern pastoral areas of Ethiopia indicated that farmers who have access to market information, irrigation, machinery and who own several farm plots are more likely to diversify. These preconditions point to the suitability of this tool in practice.

As far as the suitability of crops or livestock is concerned, certain strategically chosen combinations, on the basis of their varying degree of resistance to the main elements of hazard such as pests, diseases, drought conditions, and market potential among others, are more effective. For example:

- Sorghum and maize may be grown together because sorghum is drought resistant but susceptible to bird damage, whereas maize is liable to fail in a drought but is more resistant to bird damage.
- Mixed broadcast farming traditionally practiced in low rainfall areas of Southern Africa, such as the intercropping of maize, sorghum, and pumpkin.
- Crop diversification including grains and nitrogen-fixing legumes either in a rotation or mixed/intercropping.
- Maize and beans to protect the income as their market prices are likely not correlated with each other.

3. Advantages

There are several advantages of agricultural diversification. For example:

- It reduces farm income variability (risk), potentially ensure adequate cash flow, debt obligations and family livelihood.
- It can potentially protect from different sources of farm risk such as market and climate.
- It can exploit the potential symbiotic/synergetic relationship between different production activities such as crops, livestock and fisheries;
- The diversification of production by individual farmers may not necessarily imply more choice for consumers, although, farmers are also consumers and self-consumption is also a risk management strategy. Moreover, collectively with the introduction of various niche and specialty products, there could be more diversified produce available in the market.

4. Disadvantages

FAO (2008) and ICRISAT publications list several negative factors and their potential impact. Thus, the following possible limitations can occur:

- If the yields of two commodities have a strong positive correlation, the aggregate risk may not be reduced;
- Diversification may result in increased capital requirements for additional enterprises;
- A highly complex diversification plan may increase risk as a result of the need for managing new enterprises considering the workload bottlenecks, competition over limited land and capital resources;
- Adding new niche crops/products may carry more risks due to the requirements of new production techniques, marketing methods, skills, training, new ways of reaching markets, etc.

Some other mitigating measures proposed in other tools of ARM such as agricultural insurance, contract farming, marketing arrangements, etc. can be combined with this tool to counteract the influence of these factors.

5. Examples/Case studies

CASE STUDY 2: Diversification into dairy – India's milk revolution (World Bank)

This case study deals with a strategy implemented successfully by the Government of India on a large scale that has led to diversified farming including dairy production, especially by marginal and smallholder farmers in the country. The strategy was based on the principles of decentralized production and centralized processing and marketing with village level cooperatives and central level federations which are run democratically. Before the implementation of this scheme, the country was faced with constraints such as the scarcity of milk, farmers without land and livestock, facing crop failures and not profiting from milk production due to middle men. With the successful implementation of the Operation, the farmers who have participated in the program are now better informed about the technologies and economics of efficient milk production, most own highly productive cows and buffaloes and earn a steady income. This case study illustrates how this tool can be effective in managing risk at the farm, community and national levels.

India's Milk Revolution: Investing in Rural Producer Organizations

Over the last 25 years or so, the Indian dairy industry has progressed from a situation of scarcity to one of plenty. Dairy farmers today are better informed about technologies and economics of efficient milk production. Even landless and marginal farmers now own highly productive cows and buffaloes in many areas. Modern technology and advanced management systems in milk processing and marketing have brought about a marked change in the marketplace. Consumers now have a wide range of choice of products and packages. The Operation Flood (OF) program implemented by National Dairy Development Board (NDDB) played a key role in this transformation. The importance of the OF program lies in its focus on small rural producers. Lucrative alternate employment opportunities are often not available in Indian villages, making dairying an attractive option. Low capital intensity, a short operating cycle, and steady returns make dairying a preferred activity among marginal and small farmers (those having less than two hectares of land) and even for the landless, who depend on common grazing and forest lands for fodder. Nearly 70 million households own a total of 98 million cows and buffaloes. A majority of milk producers have 1–2 milk animals and account for some 70 percent of the country's milk production. On an average, about 22.5 percent of the income of the rural households is contributed by milk.

Anand-pattern dairy cooperatives

Anand-pattern cooperative structures are comprised of village-level dairy cooperative societies (DCSs). These promote district-level unions, which in turn promote a state-level marketing federation. Starting in 1970, NDDB replicated the Anand-pattern cooperatives through the OF program all over India. The Anand pattern envisaged:

- Decentralized milk production by small milk producers
- Milk procurement by primary dairy cooperatives of milk producers
- Centralized milk processing by union of dairy cooperatives
- Marketing of milk and milk products by a federation of unions.

The primary milk producers democratically govern the entire federal cooperative structure to ensure that the higher tier organizations are geared to serve the purpose of the lower. This ensures that the gains at all levels ultimately flow back to the milk producers in significant measure. The core feature of the Anand pattern is farmer control at all three stages—procurement, processing, and marketing—of milk and milk products.

Impact and results

Over the last 30 years or so, the Indian dairy industry has progressed from a situation of scarcity to one of plenty, making India the largest producer of milk in the world. The credit for this should mainly go to the OF program. Milk production in 1968–69, just before the launching of Operation Flood, was only 21.2 million metric tons. It increased to 30.4 million metric tons by 1979–80, 51.4 million by 1989–90, and 84.6 million by 2001–02. As a result of substantial increase in milk production, milk consumption in India has risen from a low of 107 grams per day in 1970 to more than 226 grams per day in 2002.

The OF program was funded by a World Bank loan, European food aid, and internal resources of NDDB. Total investments at the end of Phase III were estimated at Rs 15.87 billion. By 1996, the higher growth rate attributed to the OF program resulted in an extra 43 million metric tons of milk per annum. Since the start of this accelerated growth trend, the total increment has been 1,086 million metric tons. Each ton would require about $310 of imported ingredients if it were to be replaced with recombined milk. If even 2 percent of the observed increase in milk production were due to all investments from World Bank, the European Union, and NDDB's own resources, it would return an economic rate of return (ERR) of 10 percent. The returns are phenomenal if most of the increased growth is attributed to OF program. Partly this is due to the congenial environment created by OF.

...

Implementation factors

The success of a mammoth program like OF requires meticulous planning and implementation based various principles and factors. First, OF cooperatives have created a grassroots foundation underpinning India's democracy. Second, the instrument of development—dairy cooperatives—is entirely in the hands of the farmers who benefit from them. Third, the OF program sought to establish milk producers' cooperatives in the villages with the broad objective of increasing milk production ("a flood of milk"), augment rural incomes, and transfer to milk producers the profits of milk marketing, which hitherto had been enjoyed by well-to-do middlemen. As the farmers progressed through the learning process, other cooperatives were formed and brought within the organizational umbrella. Gradually methods were refined, and the organization that was eventually to become the NDDB grew—from the bottom up—adding new layers and branches as it grew.

Lessons and issues

Several propositions are embodied in the core of the design concept underlying the Anand pattern. Violation of these could explain numerous failures not only in attempted replication of the Anand pattern but throughout the arena of development experimentation.

- *Market access, the pre-condition for post-subsistence production.* In a subsistence production system, raising production and productivity requires that subsistence producers have easy, low-cost access to a stimulated and expanding market. Such access helps the producers use up the slack available in their production systems.
- *Marketing, the first step to cooperative organization.* To mount a successful marketing strategy it is best to begin by studying demand rather than the production system. Where marketing is underemphasized or mishandled, dairy and other cooperatives fail.
- *Anand's superior design concept.* A superior design concept is required to avoid a mismatch between demand and supply variations and to free the cooperative from cutthroat competition with small-time players.
- *The principle of pump priming.* The best way to organize a producers' cooperative is to start with marketing. However, unless producers' cooperatives are organized, they have nothing to market, and unless cooperatives know how to dispose of their produce, they cannot start the procurement process. NDDB found pump-priming the best answer to the launch problem that all new co-operatives face.
- *Member control and professional management.* In the absence of professional expertise, it would be difficult to quickly gain a sufficient market foothold and to exploit the full advantage offered by technology and market. The interests served by a successful business enterprise will depend upon whom professionals are accountable to, in principle and in practice.

Source: World Bank. 2004. *A case study from reducing poverty, sustaining growth – what works, what doesn't and why;* A global exchange for scaling-up success. A global learning process and conference, Shanghai; May 25-2, pp. 49-52
http://web.worldbank.org/archive/website00819C/WEB/PDF/SCALING_.PDF

6. Assessment Test for Tool # 2: Crop and Enterprise Diversification

Indicate whether each statement below is true or false.

1. Many farmers around the world, especially smallholder farmers, practice diversification by integrating crops and livestock to reduce risk and to improve their efficiency in resource use and the sustainability of the natural resource base.

2. The basic principle for formulating an effective and efficient diversification plan is that the earnings of various production activities (based on production and net prices) are strongly positively correlated with each other.

3. Most small holder farmers are *risk takers since they have to deal with everyday risk in agriculture. This is important because* in order to decide which specific plan among the set of efficient plans is the right one for a given farmer, one has to look at the farmer's risk preference.

4. Credit availability and proper infrastructure for post-harvest and marketing activities are some of the determining factors to enable farmers to diversify towards higher-value production.

5. Farmers will not diversify their production due to changing government policies like subsidies/taxes on different commodities.

6. Farm diversification provides a richer choice to consumers, potentially helping to improve the nutrition of the population being served.

Answers

1. True
2. False. They should be negatively or weakly positively correlated so that the total variance including the covariance can be lowered.
3. False. Most small holder farmers are risk averse.
4. True
5. False. Most likely they will.
6. True

7. References

1. **Blank, Steven C.** 1990. Returns to Limited Crop Diversification. *Western Journal of Agricultural Economics,* 15(2): 204-212.

2. **FAO.** 2015, Aug. *Livelihood diversification and vulnerability to poverty in rural Malawi,* by Asfaw S., McCarthy, N., Paolantonio, A., Cavatassi, R., Amare, M., Lipper, L. ESA Working Paper No. 15-02 http://www.fao.org/3/a-i4887e.pdf

3. **FAO.** 2013. *A policymakers' guide to crop diversification: The case of the potato in Kenya.* http://www.fao.org/docrep/018/i3329e/i3329e.pdf

4. **FAO.** 2012, May. *FAO Diversification booklet series.* *http://www.fao.org/ag/ags/newsandevents/news/en/c/143672/*

5. **FAO.** 2011. Products and profits from poultry. Diversification booklet number 3, 2nd Edition, by Hilmi, Martin, Dolberg, Frands and Alders, Robyn. Rome, Italy. http://www.fao.org/3/a-i2465e.pdf

6. **FAO. 2008.** Managing Risk In Farming by David Kahan. Farm Management Extension Guide 3.

7. **FAO.** 2004, Feb. *Agricultural diversification: opportunities and constraints,* by Pingali, P. FAO Rice Conference. Rome, Italy. http://www.fao.org/rice2004/en/pdf/pingali.pdf

8. **FAO.** 2002, Dec. *Livelihood diversification and enterprise development*, by Warren, P, LSP Working Paper 4. ftp://ftp.fao.org/docrep/fao/008/j2816e/j2816e00.pdf

9. **ICRISAT.** *Diversification. Webpage.* http://exploreit.icrisat.org/html2pdf_v4.03/examples/pdf.php?page=196

10. **ICRISAT.** Dec. 2013. Exploring intensification options for mixed crop-livestock farms in southern Mali. http://www.icrisat.org/locations/wca/ar-ICRISAT-2013.pdf

11. **IFAD.** 2012, Mar. *Farm intensification, crop diversification and non-farming jobs,* Independent evaluation, No.18. http://www.ifad.org/evaluation/public_html/eksyst/doc/insight/pf/rwanda12.pdf

12. **Little, P. D., Smith, K., Cellarius, B. A., Coppock, D. L. and Barrett, C.** 2001. *Avoiding Disaster: Diversification and Risk Management among East African Herders. Development and Change*, 32: 401–433.

13. ***Markowitz, H.M.*** *1959. Portfolio Selection: Efficient Diversification of Investments. New York: John Wiley & Sons.* (reprinted by Yale University Press, 1970; 2nd ed. Basil Blackwell, 1991)

14. **Mesfin, Wondimagegn.** 2011. Pattern, Trend and Determinants of Crop Diversification: Empirical Evidence from Smallholders in Eastern Ethiopia. Journal of Economics and Sustainable Development. Vol.2, No.8, pp. 78-89.

15. **OECD**, 2009. Managing risk in agriculture: a holistic approach (*extracts*). Paris.

16. **World Bank, FAO and IFAD**. 2008. Gender in Crop Agriculture Sourcebook. Module 12. http://siteresources.worldbank.org/INTGENAGRLIVSOUBOOK/Resources/CompleteBook.pdf

17. **World Bank.** 2005. Managing Agricultural Production Risk Innovations in Developing Countries. http://siteresources.worldbank.org/INTARD/Resources/Managing_Ag_Risk_FINAL.pdf

18. **World Bank.** 2004. *A case study from reducing poverty, sustaining growth – what works, what doesn't and why;* A global exchange for scaling-up success. A global learning process and conference, Shanghai. http://web.worldbank.org/archive/website00819C/WEB/PDF/SCALING_.PDF

Tool # 3: Assets and Income Based Strategies

1. What is it? – Concepts and Characteristics

Farm family assets or wealth, including income, are traditionally used to cushion the family's welfare from shocks to the farming business and the livelihood derived from it. This tool focuses on how income earning alternatives and asset allocation at the farm-level, as well as at the regional-level can be used to reduce and mitigate the risk in agriculture.

Some of the traditional methods of **asset diversification** include maintaining a balance between productive assets such as land, livestock, irrigation systems, machinery, food stocks, etc. and liquid or near liquid assets such as savings account, jewellery, food reserves, small animals, etc. **Income diversification**, on the other hand is a much more active strategy where the objective is to maintain the continuity of income flow through alternative farm and non-farm sources.

This strategy of asset and income diversification can be applied at the farm level by an individual farmer for his/her own farm/family business. To help all or at least a large number of farmers in a given area, this strategy requires the creation of off-farm income earning opportunities such as further processing, agribusinesses and non-agricultural/industrial activities. This can be seen as spatial diversification, so that a single natural disaster, such as drought, does not wipe out the whole area. A well-diversified local rural economy makes the rural inhabitants, including farmers, much more resilient in maintaining their livelihoods.

Income diversification for a farm family is an extension of crop or enterprise diversification that includes non-farm income earning opportunities by using a part of the farmer's own labour so that the total farm family income (farm plus non-farm) is higher and more stable. The underlying theoretical considerations are the same (as outline in Tool 2, Text Box).

Having a non-farm or off-farm income as a fall back support, the farmer may be willing to make investments on the farm that he/she would otherwise not make, thus taking advantage of new and improved technologies, making the farm more productive and profitable in the long run. However, from the meso-macro point of view, it can aid in strengthening the rural economy to generate employment and incomes so that livelihood sources are diversified and farmers themselves also benefit from this to better manage their own farming risks.

The motivation for farm household members to seek non-farm work and income opportunities could be a combination of "push" factors (such as relatively poor and uncertain returns to labour in farming and the need for additional income to cover expenses, debt, etc.) and "pull" factors (e.g. attractive and relatively certain wages/returns in the non-farm sector).

Many studies in the literature (Carletto et al. 2007, Barrett et al. 2001, World Bank 2014) have shown that the rural non-farm economy (RNFE)[5] plays a critical role in generating the income of rural households and therefore in food security as well. A review by Reardon et al. (1988) showed that rural non-farm activities account for 42 percent of rural household incomes in Africa, 40 percent in Latin America and 32 percent in Asia. A study by COMESA (2009) showed that income

[5] According to Davis (2004) "the rural non-farm economy (RNFE) may be defined as comprising of all those non-agricultural activities which generate income to rural households (including income in-kind and remittances), either through waged work or self-employment." Although there can be some distinction between the off-farm and non-farm income earning activities, for practical purposes they are used interchangeably in this report.

diversification is an increasingly important means for herders to manage risk and that the proportion from non-pastoral sources exceeds 20 percent in pastoral locations in the COMESA region.

The farming sector can have backward linkages to input supplying sectors and forward linkages to output processing sectors. Strong ties among all three sectors can create more opportunities for farm labour to be used in the non-farm sector, increase the incomes of farm families and help develop the agriculture sector further. Therefore, a diversified rural economy increases the capacity to increase productivity, reduce risk and help mitigate the impact of adverse events in agriculture.

The challenge for policy makers and other stakeholders, therefore, is to develop economic diversification which can prove to be an important tool (complementary to other applicable tools) of agricultural risk management. In this regard, a publication by the World Bank (2005) suggested that policy makers should incorporate risk management strategies into rural development strategy formulation and that this "strategy should consider what reforms are needed to encourage income diversification and to allow farmers a full range of choices in a functioning marketplace." Thus, this particular tool can also be seen as a policy instrument, useful in achieving the goal of improved ARM.

There is a need and potential for public private partnership in designing and implementing programs under this tool.

2. Suitability of the tool

This tool is suitable for farm families where non-farm employment opportunities exist and the farm family has excess labour to hire out on a seasonal or regular basis.

It is useful to manage risks related to the farm production and market price (by being able to afford to wait to sell the produce at the opportune time). Having an additional non-farm income as a source of self-insurance can help the farmer exploit the use of other tools of risk management.

Typically, improved land access, higher levels of education and greater access to infrastructure appear to be most closely linked to non-agricultural wage employment (Winters et al. 2009).

In the majority of cases there seems to be a positive correlation between participation in non-agricultural activities and household welfare status. Furthermore, within non-farm activities, the results indicate, in virtually all countries, that better-off households are more likely to participate in non-agricultural wage labour (World Bank 2014).

These empirical research results reveal a challenge to governments and policy makers as to how to increase the participation of the poor and uneducated, who have no access to infrastructure, in income earning opportunities in a non-farm rural economy so that they can come out of the poverty trap and also manage risk with positive results in their own farming activities. (Barrett et al. 2001)

3. Advantages

There are several reasons why the promotion of income diversification, especially through RNFE activity, can be of great interest to policy-makers in developing countries (FAO 1998). They are:

- Evidence shows that non-farm income is an important factor in food security, since it allows greater access to food.

- It can potentially protect from different sources of farm household risk such as market, climate or farming risk.
- This source of rural income may also slow down the pace of urbanization as well as natural resource degradation through overexploitation.
- Non-farm income can relieve credit constraint, positively affecting the performance of agriculture by providing farmers with cash to invest in productivity-enhancing inputs, technology and infrastructure.
- The development of RNFE activity in the food system (including the agro-processing industry, the distribution/marketing of output and the provision of farm inputs) may increase the profitability of farming by increasing the availability of inputs and improving access to market outlets.
- Conversely, in a symbiotic relationship, strong and dynamic agriculture can help the non-farm sector to flourish as farm output is available for processing and distribution where there are inputs to be sold and equipment repaired and where farm cash incomes are spent on local goods and services.
- A village-level World Bank (2013) study in India documents that "the non-farm sector is not only increasing incomes and reducing poverty, but appears as well to be breaking down long-standing barriers to mobility among the poorest segments of rural society," concluding that the process of diversification could yield significant returns in terms of declining poverty and increasing income mobility.
- According to FAO (2012), diversification in the food production system can bring wider benefits to society through:
    - Food security, improved nutrition and health,
    - A secure source of income, employment and high value production, and
    - The resilience of farming systems and environmental services.

4. Disadvantages

- It is possible that there is a negative correlation between agricultural diversification and the levels of efficiency (As found in a study from Cameroon by Nzie et al. 2012 for producers of plantain). In the study, producers in some regions, who did not diversify their activities, turned out to be more efficient - implying that specialization lead to higher levels of productivity and incomes.

In addition, as described by several researchers, (e.g. Barrett et al. 2001, Abdulai et al. 2001, Woldenhanna & Oskam 2001, and Ibekwe et al. 2010) the literature points to the following limitations (which can also be seen as challenges):

- Substantial entry or mobility barriers to higher return opportunities seem to exist within the rural non-farm economy, leaving the poor with less diversified assets and income portfolios to bear both lower expected returns and a higher variability in earnings. Furthermore, Woldenhanna & Oskam (2001) suggest that due to entry barriers in Ethiopia, relatively wealthy farm households may dominate the most lucrative rural non-farm activities such as masonry, carpentry and petty trade.
- Opportunities for income diversification are generally weaker to those with no physical access to the market.
- Attractive non-farm opportunities are accessible only to a select few educated and skilled among the many poor in rural Africa.
- Canagarajaha et al. (2001) found that there is a tendency of non-farm income to contribute to inequality, especially among female-headed households. Furthermore, a World Bank

study (2013) showed that a significant increase in income inequality accompanied diversification in a rural non-farm economy.

5. Examples/Case studies

EXAMPLE: The Magnitude Of Farm/Non-Farm Linkages

On the basis of state- and district-level data for rural areas, rural towns and the combined area in India, Hazell and Haggblade[1] found that on average a 100 rupee (Rs) increase in agricultural income is associated with a Rs 64 increase in RNF income, distributed with Rs 25 in rural areas and Rs 39 in rural towns. Infrastructure, rural population density and farm income levels increase the multiplier. Thus, the figure is as high as 93 in states characterized by high agricultural productivity, high rural population density and rur-urbanization, such as Punjab and Haryana, but only 46 in low productivity states (such as Bihar).

The IFPRI Annual Report 1985 shows that, in North Arcot district in the Indian State of Tamil Nadu, a 1 percent increase in agricultural output is associated with an additional 0.9 percent growth in non-farm employment. Also from North Arcot district, Hazell, Ramasamy and Rajagopalan[2] found (using 1982/83 data) that a Rs 1 increase in agricultural value added generated Rs 0.87 of additional value added in the non-farm sector.

Bell, Hazell and Slade[3] found that, in the Muda River region of Malaysia, an increase of agricultural income of 1 percent induced an additional increase in other rural income of 0.83 percent.

Using data from Sierra Leone and Nigeria, Haggblade, Hazell and Brown[4] find multipliers in the order of 1.5; hence a $1 increase in agricultural value added in those African countries generated an additional $0.5 of rural income which is lower than the figures from Asia quoted above.

The African multiplier was generated in a proportion of about 80 percent by expenditure (as opposed to production) linkages, while in the Asian cases the expenditure linkage effect is a lesser share of the total: in the Muda case with which they contrast it, consumption linkages account for only 60 percent of the total multiplier, and in the North Arcot case, only 50 percent.

1 P. Hazell and S. Haggblade. 1991. Rural-urban growth linkages in India. India Journal of Agricultural Economics, 46(4): 515-529.
2 P. Hazell, C. Ramasamy and V. Rajagopalan. 1991. An analysis of the indirect effects of agricultural growth on the regional economy. In P. Hazell and C. Ramasamy, eds. The green revolution reconsidered: the impact of high-yielding rice varieties in South India. Baltimore, USA, The Johns Hopkins University Press.
3 C. Bell, P. Hazell and R. Slade. 1982. Project evaluation in regional perspective: a study of an irrigation project in northwest Malaysia. Baltimore, USA, The Johns Hopkins University Press.
4 S. Haggblade, P. Hazell and J. Brown. 1989. Farm-nonfarm linkages in rural sub-Saharan Africa. World Development, 17(8): 1173-1201.

CASE STUDY 3: Assistance to small-scale food processing enterprises Tanzania (FAO)

The following case study deals with the implementation of an income diversification program through the provision of a variety of resources aimed at women and others in poverty (the main stakeholders) in the United Republic of Tanzania. Some of these resources include food processing and entrepreneurship development courses, opportunities for women in business to sell their products, consultancy and networking as well as facilities for product testing. In the case study, the tool has been implemented by the Women Entrepreneurship Development (WED) programme operating within the Small Industries Development Organization (SIDO). This tool shows how women-led families can benefit from an income diversification program which will help them manage risks in their farm production activities.

E STUDY 4 **Assistance to small-scale food processing enterprises from a support agency**

2004, the Women Entrepreneurship Development (WED) programme in processing has operated as an independent unit within the Small Industries opment Organization (SIDO) in the United Republic of Tanzania. The programme e objective of promoting productive employment and gender equality, within the on poverty alleviation and sustainable livelihoods. The impact of this programme eached all 21 regions of the Tanzanian mainland and the islands and current amme activities include:

ood processing and entrepreneurship development courses including Training f Trainers courses. WED has modern facilities for training entrepreneurs and ainers in food processing methods and technologies, food quality and safety, omplemented by business and entrepreneurship skills for managing enterprises. he distinctive features of the training are a combination of entrepreneurship nd technical skills offered by experienced trainers with reputable credentials. raining is tailor-made according to the needs of course participants, using articipatory training techniques and sharing experiences through learning by oing. Trained entrepreneurs are equipped to start their own food processing nterprises supported by advisory services.
onsultancy for the small-scale food industry. WED offers services to food rocessing entrepreneurs in surfing the Internet, local and international technology uppliers, compiling and offering information on food processing technologies.
rovision of information on food industry markets. Finding market information n the Internet, using local and international publications, newsletters, bulletins, ooklets, etc. Preparation of feasibility and other technical studies for micro-, mall- and medium-scale enterprises (MSMEs). Business cash-flow preparation nd proposal writing.
rganizing open days and trade fairs for MSMEs to promote products made y women entrepreneurs. The programme organizes open days/exhibition for nterprises in food processing, packaging and related industries. Exhibitions rovide a forum for interactions with consumers, policy-makers, financiers and ther stakeholders; in addition to promoting manufactured products. Participating xhibitors obtain good contacts including orders from supermarkets and hotels. he programme also facilitates participation in national and sub-regional trade airs (for example Nairobi and Kampala international trade fairs).
ourcing and sale of food processing equipment and other inputs such as ackaging materials.
acilities for food processing trials and product testing. These include fruit, egetable and meat processing, pasta making, baking, mixing, milling flour and rinding spices, and oil processing. The unit also has a quality control laboratory nd water treatment plant and can be used by micro-scale entrepreneurs wishing expand their businesses.
etworking and advocacy.

WED has trained 118 trainers to date who are working in all regions of the United Republic of Tanzania and Zanzibar, with all 21 regions having 1 to 2 trainers, some located in the rural areas. In summary:

- 2 400 entrepreneurs have been trained in short courses (from 2 days to 2 months), of which 585 are 1 to 2 month courses.
- Over 70 percent of trainees are in business.
- Assets of the women in business range from US$670 to US$16 200
- 1 700 jobs have been created by both long and short course participants.
- Value added to processed foods compared to raw materials from farms is over 200 percent.
- Most of the enterprises produce quality products, which are sold in different supermarkets in the country competing with imported food products.
- Formation of a Tanzanian Food Processors Association operating independently from the programme and with over 200 paid up members is another positive impact of the programme.

Source: As cited in **FAO 2011.** Processing for prosperity. Diversification booklet number 5, 2nd edition, by Peter Fellows. Rome, Italy. Pages 70-71. http://www.fao.org/3/a-i2468e.pdf

6. Assessment Test for Tool # 3: Assets and Income Based Strategies

A South African farmer is looking to increase his household income. The list below describes several actions the farmer can take in order to increase his household income and crop yield. Match the description of each action to the type of diversification it corresponds to. One letter may apply to more than one action.

1. Intercropping sorghum, drought resistant but susceptible to bird damage and maize, more resistant to bird damage but susceptible to drought.
2. The farmer's wife works as a nurse while the farmer works on their farm.
3. Both cowpea and maize can be used as feedstuff for livestock. The interaction between the two provides a supply of nitrogen in the soil. Maize is used as a staple food but can also be further processed into cottage industry products such as popcorn, oil, adhesive, etc.
4. The farmer decides to cultivate maize during the months of November to February, and takes up a labour job on another farm while renting out his land for wind turbines.

A. Crop diversification
B. Enterprise diversification
C. Income diversification

Answers

1. A
2. C
3. B, or C (if carried out off-farm and under different management)
4. C

7. References:

1. **Abdulai, A., CroleRees, A.** 2001, Aug. *Determinants of income diversification amongst rural households in Southern Mali,* Food Policy 26(4) pp.437-452.

2. **Barrett, C. B., Reardon, T., Webb, P.** 2001, Aug. *Nonfarm income diversification and household livelihood strategies in rural Africa: concepts, dynamics, and policy implications,* Food Policy 26(4) pp.315-331

3. **Canagarajaha, S., Newman, C., Bhattamishra, R.** 2001, Aug. *Non-farm income, gender, and inequality: evidence from rural Ghana and Uganda,* Food Policy 26(4) pp.405-420.

4. **Carletto, G., Covarrubias, K., Davis, B., Krausova, M., Stamoulis, K., Winters, P., & Zezza, A.** 2007. *Rural Income Generating Activities in Developing* Countries, Journal of Agricultural and Development Economics, 4(1), 146-193.

5. **COMESA** (Common Market for Eastern and Southern Africa). 2009. *Income Diversification Among East African Pastoralists: Lessons for policy makers.* Policy Brief No. 3

6. **Davis, J.** 2004. *The Rural Non-farm Economy, Livelihoods and their Diversification: Issues and Options.* Chatham, UK: Natural Resource Institute.

7. **FAO.** 2012, May. *FAO Diversification booklet series.* http://www.fao.org/ag/ags/newsandevents/news/en/c/143672/

8. **FAO** 2011. Processing for prosperity. Diversification booklet number 5, 2nd edition, by Peter Fellows. Rome, Italy. http://www.fao.org/3/a-i2468e.pdf

9. **FAO.** 2002. *Promoting farm/non-farm linkages for rural development: Case studies from Africa and Latin America.* Rome, Italy.

10. **FAO.** 1998. *Rural Non-Farm Income in Developing Countries. The State of Food and Agriculture (SOFA) 1998.* Rome. http://www.fao.org/docrep/017/w9500e/w9500e.pdf

11. **Ibekwe, U.C., Eze, C.C., Ohajianya, D.O., Orebiyi, J.S., Onyemauwa, C. S., Korie, O.C.** 2010. *Determinants of non-farm income among farm households in South East Nigeria,* Researcher 2(7) pp. 49-52.

12. **OECD.** 2000. *Income risk management in agriculture,* Agriculture and Food.

13. **Reardon, T., Stamoulis, K., Cruz, M-E., Balisacan, A., Berdegue, J. and Banks, B.** 1998. *Rural Non-Farm Income in Developing Countries. The State of Food and Agriculture 1998: Part III.* Rome: Food and Agriculture Organization of the United Nations.

14. **Siegel, P. B.** 2005. "Looking at Rural Risk Management Using an Asset-Based Approach." Paper prepared for the Commodity Risk Management Group, Agricultural and Rural Development Department, ESW, The World Bank, Washington, D.C.

15. **Winters, P., Davis, B., Carletto, G.** 2009, Sep. *Assets, activities and rural income generation: Evidence from a multicountry analysis,* World Development Vol. 37, pp. 1435-1452.

16. **Woldenhanna, T., Oskam, A.** 2001, Aug. *Income diversification and entry barriers: evidence from the Tigray region of northern Ethiopia,* Food Policy 26(4) pp.351-365.

17. **World Bank.** 2014, Nov. *Income diversification patterns in rural Sub-Saharan Africa: Reassessing the evidence,* by Davis, B., Giuseppe, S. D., Zezza, A. Policy Research Working Paper 7108

18. **World Bank** 2013, May. *Non-farm diversification, poverty, economic mobility and income inequality: a case study in village India,* by Himanshu, Lanjouw, P., Murgai, R., Stern, N. Policy Research Paper 6451.

19. **World Bank.** 2005. *Managing Agricultural Production Risk Innovations in Developing Countries*. http://siteresources.worldbank.org/INTARD/Resources/Managing_Ag_Risk_FINAL.pdf

LESSON 2
Finance Related Risk Management Tools

Learning objectives

At the end of the lesson, learners should be able to understand and apply various finance related risk management tools, appreciate the advantages/benefits and disadvantages/costs of the selected tools, their complementarities with other tools, and to judge their suitability in a given area or farm situation.

Various finance related instruments are grouped into three sets of tools, namely:

Tool 4: Agricultural Insurance
Tool 5: Weather Index Insurance, and
Tool 6: Agricultural Finance and Microfinance

Not all tools will be applicable to every farm or locality but there may be some that are suitable and accessible in a specific situation. From the point of view of other stakeholders, including policy makers, an understanding of all available modern tools will be helpful.

Historically, countries have tried to manage the outcome of commodity problems by manipulating and controlling demand and supply. Experiences from developed and developing countries have shown that this way of managing risk (especially price and net income risk) in agriculture is inefficient, ineffective and very costly. Hence, modern solutions have aimed for more market orientation, with the involvement of farmers, government, national/international donors and private sector actors dealing with financial and market sector players. In addition to traditional agricultural credit and insurance institutions, it has led to the development of innovative risk financing instruments such as microfinance institutions, weather index insurance schemes, microinsurance, etc.

These financial tools are based on the principle of risk sharing. Agricultural insurance systems, for example, pool risks from a large number of clients and share it with wider national or international financing firms. The schemes are possible where they are large enough to pool the risk successfully and where they are actuarially sound. They provide protection to an individual farmer in the form of potential compensation for losses in exchange for premiums paid upfront by or on behalf of the farmers.

There are probably several different financial instruments available in different countries, but the three main instruments are described here, as the most applicable ones to agricultural risk in Africa.

Finally, it should be mentioned that this lesson does not cover all possible tools. For example, various formal and government financing schemes, lines of credit from a bank, use of small livestock as near liquid assets, financing through relatives or remittances from family members, among others, are not covered.

Key words for lesson 2:

1. *adverse selection,*
2. *correlated risks,*
3. *crop insurance,*
4. *group loan,*

5. *livestock insurance,*
6. *loan defaults,*
7. *microcredit loans*
8. *moral hazard,*
9. *payment levels,*
10. *policy holders,*
11. *rural microfinance,*
12. *systemic risk*
13. *transaction cost,*
14. *trigger level,*
15. *weather index insurance*

Tool # 4: Agricultural Insurance

1. What is it? – Concepts and Characteristics

The primary objective of this lesson is to provide a basic understanding of different types of insurance schemes, their applicability and the pros and cons of using insurance as a risk sharing tool.

Insurance is a mechanism used to transfer a specified risk by the insured to a third party; normally an insurer. The insured pays a premium to the insurer and in exchange, the insurer agrees to pay an indemnity (compensation for suffered, assessed losses) that occurs during and in accordance to the period as well as the terms and conditions of the insurance policy. Agricultural insurance is an ex-ante measure to cope with farm production and/or revenue losses, thus helping to smoothen farm income over the years.

Crop insurance to protect a farmer from unforeseen unfavourable shocks on yield or production can be a very useful tool and is more commonly used in developed countries. However, the application of this product in developing economies is very low and has proven complex. The results of efforts undertaken have borne mixed results. Recent assessments show difficult challenges in improving insurance acceptance.

As opposed to index-based insurance, specific yield or production insurance is generally for individual farmers as policy holders with their own contract covering protection against their own stated perils. Index-based insurance[6], on the other hand, relies on the estimated losses based on the index value and does not require a field assessment of each policy holder. Insurance offered primarily to clients with a low income and limited access to mainstream insurance services and other means of effectively coping with risk is called microinsurance. Microinsurance differs from traditional insurance as it requires targeting people with a low income and catering for their specific needs such as affordability and inclusiveness, simplicity and clarity in documentation, accessible processes, and building trust among target clients (McCord/IFAD 2012).

The **global usage** of agricultural insurance, measured by the agricultural premium volume, increased dramatically between 2004 and 2007, rising from $8 billion to about $20 billion, primarily due to rising agricultural commodity prices (World Bank 2010b). The same World Bank survey also showed that livestock insurance is available in 85 percent of the 65 surveyed countries. The insurance coverage measured by the premium volume is found to be much lower for livestock insurance than for crop insurance.

Crop insurance is by far the most popular risk management tool used by U.S. crop producers, covering more than 200 million acres with a total premium of over $3 billion in 2003 with the two most popular crop insurance products against low yields and against revenue drops (Babcock et al. 2004). According to the USDA (2015), starting in the 2016 crop year, a new risk management tool called a Margin Protection Plan will be available in the United States in addition to traditional crop insurance policies in select counties starting with corn, rice, soybeans, and spring wheat. Under this insurance policy, farmers can protect their per acre net operating income, (i.e. a margin which is equal to expected revenue minus non-land costs) thus covering risks of unfavourable changes in yield, price and costs at a specified level.

---

[6] Given the prominence of Weather Index Insurance within the general instruments of agricultural insurance, it has been covered as a tool on its own (see Tool 5).

According to the USDA's Risk Management Agency (RMA), under **crop revenue insurance,** compensation is paid to the insured farmer based on gross revenue shortfalls instead of just yield or price shortfalls. This insurance is subsidized and reinsured by the RMA. Under the **crop yield insurance** program, indemnities are paid to producers when yields fall below the producer's insured yield level due to natural causes. Crop yield insurance is also subsidized by the USDA's Risk Management Agency. The budgetary costs of insurance subsidies in the US are very high and controversial.

Agricultural insurance is primarily used to insure production, and in some cases revenue. For managing price and market risks, other market based tools, described in the next lesson, are more appropriate.

In developing countries many pilot programmes have been developed over the years, targeting especially small-scale farmers. However, the use of agricultural insurance is still extremely low as it primarily involves farmers in developed countries (FAO 2005). Although the continent has 17 percent of the world's pastures and arable land, the value of premiums for agricultural insurance in **Africa** represents less than 0.7 percent of the world's total (Centre for Financial Inclusion 2014).

In recent years, however, with new and innovative programs such as index-based insurance aimed at small farm holders, the popularity of crop and livestock insurance is on the rise. According to the MicroInsurance Centre, in 2014 a total of 1.1 million people in the agriculture sector (including government subsidised insurance) in Africa were insured. Agricultural insurance has been one of the highest growth sectors (McCord 2015). In **Kenya,** agricultural insurance has staged a comeback in the last ten years, particularly due to the commercialisation of the dairy sector which increased the demand for livestock insurance (Kerer 2013).

There is a need and potential for public private partnership in designing and implementing programs under this tool.

2. Suitability of the tool

This tool is suitable for all farmers especially those who face occasional catastrophic drops in production or market prices. Insurance options are more suitable for cash crops or market oriented livestock production systems.

Agricultural insurance is applicable to cover risks of a single or multiple perils not under control of the farm operator, such as droughts, floods, hail, storms, hurricanes, fire, and theft, etc. These can affect crops (area and yield), animals and farm structures. The insurance can cover production risks and, sometimes are combined with price to insure also farm income, typically for commodities and markets where a well-functioning commodity exchange exists.

Agricultural insurance, however, is not the universal solution to the risks and uncertainties that farmers face. It can only address part of the losses resulting from some perils and is not a substitute for good on-farm risk-management techniques, sound production and farm management practices and investments in technology (FAO 2005). It should be promoted only when basic agricultural services, such as the timely availability of inputs, extension services, and efficient marketing channels for agricultural outputs, are in place (World Bank 2010b).

3. Advantages

When used as part of a comprehensive agricultural risk management framework, agricultural insurance can contribute to the modernisation of agriculture (WB 2010b). With affordable insurance, poor producers can protect their assets and productive capacity and weather the shocks of disastrous occurrences better than those without. They can also correct the typical under-investment behaviour that would normally keep them in a low income or perpetual subsistence situation. In general, households more susceptible to farming risks are less likely to invest in technology, fertilizer or improved seeds for crop production. Thus, agricultural insurance products, especially those that cost less to administer (such as index-based insurance) can be seen as a preferred public policy intervention instrument particularly when combined with social safety nets to bring people out of the poverty trap (Barnett et al. 2007).

The price of the insurance policy is a critical factor in a household's decision making. For example, Cole et al. (2010) found that the demand for insurance is significantly price sensitive, with a price elasticity between -0.66 and -0.88. Thus, subsidies on premium can lead to an increased demand for insurance products. However, a note of caution by the World Bank (2010b) states, "Subsidizing insurance premiums can be justified to correct and promote competitive private agricultural insurance markets, usually in the form of public goods that enhance the risk market infrastructure (for example, data collection and management, research and development, and legal and regulatory framework)".

4. Disadvantages

- Most of the agricultural insurance products designed for an individual policy holder require pre-disaster crop inspection and post-disaster physical loss inspection and therefore their administrative costs are generally too high for smallholder farmers.
- Linkages between banks and agricultural insurance, especially in Africa, are not well established, which does not help to make insurance cost-effective for small and medium scale farmers (World Bank 2010b).

The following are, as outlined by Mahul and Stutley (World Bank 2010b), some of the limitations which have made the uptake of private agricultural insurance slow, especially in developing countries.

- **Systemic Risk** - risk that affects a large number of economic units, such as farmers and herders simultaneously, which can generate major losses to insurers.
- **Informational Asymmetries** including adverse selection and moral hazard[7] requiring measuring risks and monitoring farmer behaviour. It is very difficult and expensive for private insurers to measure risks, collect relevant data, monitor producer behaviour, and establish and enforce underwriting guidelines. These problems are minimized in index-based insurance plans.
- **Limited Access to International Reinsurance Markets**
- **Agricultural Risk Market Infrastructure** - the lack of infrastructure (such as agricultural and weather databases and crop risk models, etc.) to support the agricultural insurance industry.
- **Low Risk Awareness** – farmers may underestimate the likelihood or severity of catastrophic events.
- **Lack of Insurance Culture** – a limited understanding of insurance benefits.

---

[7] **Adverse selection** implies where 'high risk' individuals are more likely to choose an insurance contract and more likely to have an accident and claim. And **moral hazard** occurs where the insured may have weaker incentives to avoid risky behaviour and will be less cautious and have more accidents/claims once insured.

- **Regulatory Impediments** – a lack of enabling regulatory frameworks governing insurance markets

5. Examples/Case studies:

Example:

Box: Crop Insurance Schemes in India

On 14 January 2016 Indian Government announced a new insurance program, called the Prime Minister Crop Insurance Scheme, under which farmers will pay between 1.5 and 2 % premiums for food grains and oilseeds and up to 5% for horticultural and cotton crops starting from the current season. The Government's target is to cover nearly 50% of farmers all over the country.

This new schemes follows a set of crop insurance schemes employed in the country. Agriculture Insurance Company of India (AIC) and ten private General Insurance Companies to implement crop insurance programme. The National Crop Insurance Programme (NCIP) consists of three main components:

- Modified National Agricultural Insurance Scheme (MNAIS)
- Weather Based Crop Insurance Scheme (WBCIS)
- Coconut Palm Insurance Scheme (CPIS)

All of these schemes involve various levels of subsidy in the premium rates up to 75 percent.

*Source:*http://economictimes.indiatimes.com/news/economy/policy/cabinet-approves-new-crop-insurance-scheme/articleshow/50559022.cms, and Information provided by the Minister of State for Agriculture. Press Information Bureau, Ministry of Agriculture, Government of India. 04-August-2015. http://pib.nic.in/newsite/PrintRelease.aspx?relid=124296

Case study

CASE STUDY 4: Indemnity-based crop insurance in Zambia (FARMAF)

The primary stakeholders of this case are smallholder farmers (members of the Zambia National Farmer's Union, ZNFU) who are participants in the crop insurance scheme, Lima Credit Scheme (LCS) who is the main credit provider, the District Farmers Associations (DFA) who co-guarantee the loan, the Zambia State Insurance Company (ZSIC) and other insurance and finance institutions in the country providing insurance against natural hazards such as droughts, lighting, floods, hailstorm, and fires. Judging from the significant increase in the number of farmers and the hectares covered over the last six years, this scheme can be seen as a success. This case study shows how this tool can be instrumental in increasing a farmer's ability to manage production risk.

Indemnity-based crop insurance in Zambia

The insurance case in Zambia focuses on the Lima Credit Scheme (LCS). The objectives of LCS is to provide smallholder farmers without collateral with commercial agricultural credit services based on group savings and loans approach. The District Farmers Associations (DFA) has to co-guarantee the loan. Moreover, the Agrisure policy issued by the Zambia State Insurance Company (ZSIC) is a mandatory element within LCS. In 2014, two other insurance companies have come on board namely: African Grey Insurance and Mayfair Insurance.

Farmers participating in the scheme are member of the Zambia National Farmers' Union (ZNFU). LCS funds mainly smallholders and targeted smallholder farmers average loan sizes of US$600 – US$700. Farmers are able to produce for the market (beyond subsistence) and practice farming as a (potential) business. The program targets farmers who organise themselves into groups of 10-20 farmers based on mutual trust, reputation and commodity focus.

At the start of the scheme only maize was amenable for insurance. Maize is the dominant food crop as well as cash crop in Zambia. Peril covered by the Agrisure policy includes damage or destruction of crops caused by natural events such as drought, lightning, flood, hailstorm and fire. In case of calamities the insurance indemnifies the cost of inputs for which credit was obtained. The insurance company carries out preharvest assessments. The agricultural inspector will write down the recommendations he has given to farmers with regards to improve farming practices. In case of a claim, the inspector will check the recommendations were implemented. The claim is not eligible if the agricultural recommendations are not Launched in 2008/2009 season the granted credit and thus exposure by ZSIC increased to US$ 3.98 million in the 2013/2014 season. Benefiting farmers have increased to 16,780 cultivating 36.700 hectare over the same period (Table 1). In 2008/2009 the insurance started with a premium set at 5% of the insured.

Table 1: Key characteristics of Lima credit and indemnity-based insurance scheme in Zambia.

Year	Number of farmers	Hectares	Premium (%)
2008/2009	600	600	5
2009/2010	1,334	2,229	5
2010/2011	1,511	3,320	5
2011/2012	4,723	10,300	4
2012/2013		21,000	4
2013/2014	16,780	36,700	4

Currently, the premium has been reduced to 4%. Premium differentiation to discriminate between exposure units more or less at risk is absent. The participating DFAs have paid premiums amounting US$ 423,600 under the current Lima farming season. In 2014/15 season, it's projected that farmers will pay premiums approximately US$ 625,000.

A smallholder farmer deposits 50% of the full supply of his input requirements in a fixed term collateral account. Interest payments on his deposit amounts 4%, which is lower than inflation. Input suppliers deliver on order from ZNFU to respective destinations where the DFAs management is responsible for distribution to farmers. The financing bank pays the invoice of the input supplier on confirmation of successful completion of the contract by ZNFU.

ZNFU envisages to reach ultimately 35.000 farmers in the coming years. In the Lima expansion plans ZNFU foresees that other field crops (e.g. soybean and other beans), livestock, vegetables and asset finance are incorporated into Lima. Moreover it strives to create more competitive financial service packages for small-scale farmers that not only provide access to seasonal credit but also provide access to medium & long term inputs & asset finance.

Source: **Van Asseldonk et al.** 2015. Evidence of linking crop insurance and rural credit. Policy Brief No . Page 2. http://www.farmaf.org/images/documents/Operational_materials/policy_brief/FARMAF_Policy_Brief_Insurance_2015.pdf

6. Assessment Test for Tool # 4: Agricultural Insurance

Indicate whether each statement below is true or false.

1. Agricultural insurance is a mechanism used to transfer a specified risk related to farm yield, production, price or income by the farmer (who agrees to pay a premium before the disaster), to an insurance company (who agrees to pay an indemnity after the disaster).

2. Crop insurance is one of the most popular risk management tools used by farmers in developing countries but not by farmers in developed countries.

3. Subsidies for farm insurance premiums in many developing countries may be justified for correcting and promoting competitive private agricultural insurance markets, usually in the form of public goods that enhance the risk market infrastructure (for example, data collection and management, research and development, and legal and regulatory framework).

4. The insurance industry faces the usual problems of systemic risk, adverse selection and moral hazard. However, in agricultural insurance, these problems are limited and can be overcome easily.

5. The insurance coverage measured by the premium volume is found to be much lower for livestock insurance than for crop insurance.

Answers

1. True
2. False; it is more commonly used by developed country farmers and in countries where high subsidies are available.
3. True
4. False; these problems are more evident (especially systemic risk) for agriculture.
5. True.

7. References:

1. **Babcock, Bruce A. Hart, Chad E. And Hayes, Dermot J.** 2004.Actuarial Fairness Of Crop Insurance Rates With Constant Rate Relativities. *Amer. J. Agr. Econ.* 86(3) (August 2004): 563–575.

2. **Barnett, B. J., C. B. Barrett, and J. R. Skees.** 2007. "Poverty Traps and Index-Based Risk Transfer Products." *World Development* 36 (10): 1766–85.

3. **Centre for Financial Inclusion (CFI)**, 2014.Agriculture Microinsurance in Africa: To Subsidize or Not to Subsidize? Blog *Posted by Annalisa Bianchessi, Microinsurance Network* **http://cfi-blog.org/2014/03/27/agriculture-microinsurance-in-africa-to-subsidize-or-not-to-subsidize/**

4. **Cole, S., Giné, X., Tobacman, J., Topalova, P., Townsend, R., and Vickery, J.** 2012. Barriers to Household Risk Management: Evidence from India. **IMF Working Paper** Research Department

5. **FAO.** 2011. *Agricultural Insurance in Asia and the Pacific Region,* by Stutley, Charles. Bangkok. http://www.fao.org/docrep/015/i2344e/i2344e00.pdf

6. **FAO**, 2005. Insurance of crops in developing countries. By R.A.J. Roberts. Agricultural Services Bulletin 159. Rome, Italy. ftp://ftp.fao.org/docrep/fao/008/y5996e/y5996e00.pdf

7. **Hazell, P.** 2001. Potential Role for Insurance in Managing Catastrophic Risk in Developing Countries, IFPRI Occasional Paper. http://www.agroinsurance.com/files/IDB_hazell[1].pdf

8. **IFAD,** 2009. Brief 93: Microinsurance, Trust and Economic Development: Evidence from a Randomized Natural Field Experiment. National Bureau of Economic Research Working Paper 15396 by Hongbin Cai, Yuyu Chen, Hanming Fang, and Li-An Zhou.

9. **Kerer, Jan**. 2013. Background Paper on the Situation of Agricultural Insurance in Kenya with Reference to International Best Practices. *Adaptation to Climate Change and Insurance (ACCI)* Nairobi, Kenya.

10. **Levin, T., & Reinhard, D.** 2007, Oct. *Microinsurance aspects in agriculture.* Munich Re Foundation Discussion Paper, Munich. https://www.agriskmanagementforum.org/sites/agriskmanagementforum.org/files/Documents/Microinsurance_in_Ag.pdf

11. **McCord, Michael J.** 2012. **Microinsurance product development for microfinance providers**. International Fund for Agricultural Development, Rome. http://www.microinsurancenetwork.org/microinsurance/key-concepts#sthash.pao2L8jp.dpuf

12. McCord, Michael J. and Biese, Katie, **2015. The Landscape of Microinsurance in Africa 2015 - Preliminary Briefing Note, MicroInsurance Centre.**

13. **OECD**, 2009. Risk Management In Agriculture: A Holistic Approach (*extracts*).

14. **United States Department of Agriculture** (USDA), Risk Management Agency http://www.rma.usda.gov/

15. **USDA,** 2015. Margin Protection for Corn, Rice, Soybeans and Wheat. http://www.rma.usda.gov/policies/mp/

16. **Van Asseldonk, M., Porgo, I., Hamusimbi, C., Mumba, G., Banda, J., Maitre d'Hotel, E., le Cotty, T., Burger, K., and Onumah, G.** 2015. *Is there evidence of linking crop insurance and rural credit and its potential benefits?* Policy Brief No 1, 2015. The Farm Risk Management for Africa (FARMAF) Project. http://www.farmaf.org/images/documents/Operational_materials/policy_brief/FARMAF_Policy_Brief_Insurance_2015.pdf

17. World Bank. 2012. **Mongolia - Index-Based Livestock Insurance Project: Restructuring, *Main report*.**

18. **World Bank.** 2010b. Government Support to Agricultural Insurance: Challenges and Options for Developing Countries by Olivier Mahul and Charles J. Stutley. Washington, DC.

Tool # 5: Weather Index Based Insurance

1. What is it? – Concepts and Characteristics

The weather index insurance (WII) instrument is an alternative to traditional crop insurance used in developing countries. In recent years, it has been tried mostly on a pilot project basis in many countries. The evidence so far has been mixed with some promises for upscaling and wider use as well as some cases of failure. The main objective of this tool is to provide a basic understanding of different modalities of this type of insurance, its pros and cons and to provide examples and experiences with its adoption, especially in Africa.

Index-based agricultural insurance consists mainly of the following types:

1. Area-yield index insurance which provides insurance against fluctuations (drops) in production (area and/or yield), determined on the basis of crop performance in a given area measured in terms of an index.
2. Weather index insurance based on meteorological stations data, and
3. Weather index insurance based on satellite images.
4. Index-based mortality insurance for livestock based on satellite images, and
5. Index-based mortality insurance for livestock based on animal population statistics.

Insurance schemes 2, 3 and 4 above, are focused on the performance of weather factors (perils) stipulated in the insurance contract. These have gotten more attention in recent years and thus are covered in more detail in this lesson.

Successful implementation of a WII scheme depends on keen participation and involvement/commitment of various stakeholders including -

- **Farmers** who are interested in purchasing/acquiring insurance protection (with or without subsidies),
- **Governments** to provide overall enabling regulatory and legal framework, available public weather and agronomic data, and possible subsidies or other social protection,
- **Insurance companies** willing to provide insurance,
- **Donors**, operating **INGOs** and local **NGOs** to play facilitating and catalytic roles.

WII is similar to other forms of insurance except that it follows a simplified procedure in which indemnity payments (compensation for damages or loss) are based on values obtained from an index that serves as a proxy for losses by all farmers. This eliminates the need for a costly assessment of the actual damages of individual policy holders. WII is based on the assumption that there is a strong correlation between the defined weather parameter (e.g. rainfall or temperature in a given area) and the yield of the crop(s) being cultivated by the insured farmers.

For transparency and unbiased actions required in the execution of WII, the terms of the policy contract, including the critical values of the agreed index, payoff amounts, etc., are discussed and agreed upon. The indemnity payment is based on the level of the index, gradually increasing from the predefined trigger level (say, for example, 200 mm of rain) and reaching a maximum pay-out at or below the specified level (e.g. 100 mm). The total sum of the insured amount is generally based on the cost of production and the area cultivated.

A typical contract will also specify the trigger and full payment levels of the index (for example, based on cumulative rainfall) in the key periods of risk, namely crop establishment, vegetative and crop flowering/harvesting stages, during the crop calendar. In practice, there could be other complexities related to rainfall distribution and the potential and actual impact on yield and area.

A recent World Bank/IFC study (2015) explains the nature of WII as follows: "From a behavioural perspective, index-based insurance is akin to a lottery or gambling on the weather. To sell traditional insurance, a potential customer must be convinced that there is a reasonable probability of a loss that must be covered. On the other hand, the customer only needs to believe that the weather is variable and unpredictable in order to find index-based insurance attractive. ...From an economic perspective, index-based insurance provides a buffer to protect the farmer against shocks that is similar to having savings. However, savings would be less convenient for weather protection if the farmer has not accumulated sufficient amount to cover their losses in the event of a drought".

This instrument is viable and has been adopted in developing countries where traditional crop or livestock insurance is not commonly used due to small farm holdings (acreage or herd size) and a lack of financial and insurance products/services. However, in many cases, the schemes have not been successful.

In Mongolia, a variation of WII for livestock is called Index-Based Mortality Insurance, where animal population statistics is used to determine livestock losses over a predetermined period of time (World Bank 2011b). Once losses surpass a predetermined threshold, pay-outs are triggered.

In Kenya, a strong correlation between the livestock mortality rate and weather data captured using the Normalized Difference Vegetation Index (NDVI) based on satellite data for the last 40 years on vegetation covered in the Marsabit district, was used for WII (Kerer 2013). The model predicted the actual mortality in years that were affected by drought. The basis risk, meaning the probability that the index was triggered but no actual losses occurred or that the index was not triggered but losses occurred, was below 10% which was deemed reliable enough to proceed with the commercialization of the product.

WII policies are primarily purchased by farmers, households or small-business owners but can also be distributed to them by organizations such as aid agencies, financial service providers, farmers' associations, input suppliers, processors or NGOs, sometimes bundled in with other services. For example, a bank or MFI may require a farmer to purchase an insurance policy along with a loan provided by them.

In practice, when it comes to deciding about the subsidies for such programs, most situations may require a balance of both social and commercial objectives. An insurance scheme with a clear social objective that protects lives and assets from catastrophic losses inevitably would require subsidies and special delivery channels to be aligned with relief or social protection interventions (World Bank 2011a).

Finally, it should be mentioned that there is a need and potential for public private partnership in designing and implementing programs under this tool.

2. Suitability of the tool

As a tool, WII is suitable for all farmers to transfer area, yield and production risk arising from abnormal weather such as droughts, floods or temperatures that are too high or low for crops and livestock, to an insurance provider. WII schemes can be applied to all crops. It can also be applicable to livestock with some modification, such as the use of a vegetation index (NDVI) instead of the usual rainfall index (e.g. Marsabit district in Kenya in Kerer 2013) or animal population and their mortality statistics as practiced in Mongolia. Thus, WII can be a suitable risk management tool to protect farmers/herders from animal loss based on extreme weather.

Given that the variables suitable for index should be based on a transparent process and reliable and timely data that can be verified independently, WII is more applicable to factors such as rainfall and temperature that are highly correlated with yields. Crop damage occurs when the level of rainfall is too high or too low (causing droughts or floods). Trigger levels should be identified accordingly.

Other factors such as localized hail, earthquakes, fires, etc. are not easily correlated to an observable and measurable factor and are therefore not suitable for an index-based insurance instrument. It is also possible that a simple cumulative rainfall index may be poorly correlated with crop yields. In such cases, a more sophisticated methodology involving weather yield index (WYX) models linking various rainfall derivative factors and crop yield may be more appropriate (FAO 2006).

3. Advantages

The IFAD/WFP study outlines the following advantages of WII as opposed to conventional insurance:

- **Transparency.** Index insurance contracts usually allow the policyholder direct access to the information on which the pay-outs will be calculated. Trust is strengthened by transparency.
- **No on-farm loss adjustment.** This is a primary advantage of index insurance, as on-farm estimation of losses (loss adjustment) is quite complex, costly and may not be credible in many low-income countries.
- **Lack of adverse selection.** Adverse selection occurs when potential insured parties have hidden information about their risk exposure that is not available to the insurer, who then becomes more likely to erroneously assess the risk of the insured. Traditional insurance encourages high-risk producers to insure, while risks and premiums are calculated on the average producer. Index insurance requires that all insured farmers within the defined area have the same insurance pay-out conditions, regardless of their specific risk exposure. Hence, insurers and clients benefit from reduced adverse selection.
- **Lack of moral hazard.** Moral hazard occurs when individuals engage in hidden activities that increase their exposure to risk as a result of purchasing insurance, or attempt to influence the claims outcome. These hidden activities can leave the insurer exposed to higher levels of risk than had been anticipated when premium rates were established. With WII, there is no benefit in individual producers trying to influence claims. All producers in the defined area are treated equally.
- **Addresses correlated risks.** Index products work best where there are correlated risks. Using traditional products, perils such as drought are challenging to insure.
- **Low operational and transaction costs.** Index insurance requires limited individual underwriting (client assessment). It can be distributed, and claims can be settled, at a relatively lower cost.
- **Rapid payout.** The measurement of weather station data, with no field loss adjustment, allows rapid pay-outs.

A recent study by Swain (2015), in a frequently disaster-affected state like Odisha in India, showed that "the Weather Based Crop Insurance Scheme (WBCIS, an area-based rainfall insurance) performs better than the National Agricultural Insurance Scheme (NAIS, an area-based crop yield insurance) because of its higher adoption rate, higher percentage of farmers benefited, lower premium, faster claim payment, and frequent indemnity payment."

4. Disadvantages

The same study outlines the following as disadvantages of WII:

- **Basis risk.** Basis risk in WII is a key constraint. The basis risk is the difference between the loss experienced by the farmer and the payout triggered. It could result in a farmer experiencing yield loss, but not receiving a pay-out, or in a pay-out being triggered without any loss being experienced. Index insurance works best where losses are homogeneous in the defined area and highly correlated with the indexed peril.
- **Limited perils.** WII normally covers only one, sometimes two, weather perils. Although this reduces the cost compared with multi-peril crop insurance, the product may not provide broad enough coverage to satisfy risk management needs.
- **Replication.** The triggers, limits and increments of a specific product need to be adjusted to reflect the weather parameters of each weather station. Different product designs are required for different crop types (or at least generic crop types). WII requires considerable technical work in its implementation. Also, there is a need for more weather stations in order to have an accurate picture of the local weather performance, crop losses and compensation to farmers.
- **Technical capacity and expertise are required**, particularly during the initial design phase for new products, in agro-meteorology and in operationalizing the products.
- **Lack of data.** WII depends on the availability and quality of weather data, which can drastically vary from country to country. In developing countries, the shortage of historical and real-time weather data is often a major hurdle. Weather date need to be combined with production and yield variability for a competitive index insurance policy design. Collecting good production and yield historical data at farm level is often even more challenging.

5. Examples/Case studies

CASE STUDY 5: Index-based insurance in Burkina Faso (FARMAF)

This case study is a report on index-based insurance issued by PlaNet Guarantee in Burkina Faso to provide risk protection to farmers facing potential crop damage due to drought. Microfinance institutions implement the PlaNet Guarantee, providing access to weather based insurance in selected villages. The index value is determined by monitored evapotranspiration data over three stages of crop development (30, 20 and 40 days starting after seeding). PlaNet Guarantee insurance covers maize but it aims to expand coverage to other crops, for example, cotton. The premium rates are around 10%. The Burkina Faso project is part of a larger project for West Africa, that aims to raise awareness of agricultural insurance to more than 165,000 farmers. This tool provides protection to a wide group of farmers in the geographic region, enhancing the ability of smallholder farmers to transfer their risk to the insurance providers.

Index-based insurance in Burkina Faso

In Burkina Faso, several micro finance institutions market the PlaNet Guarantee cover in 2013/2014. Confédération Paysanne du Faso (CPF) and CIRAD collaborate with PlaNet Guarantee to foster weather insurance take up in a limited number of villages and to assess the impact of insurance of farmers' income. Although the insurance contract is optional the credit agencies are becoming more stringent in requesting this cover. Insured farmers without credit are rare in Burkina Faso.

The pay-outs for index insurance relate to specific weather events which is in Burkina Faso the decadal relative evapotranspiration. The index value is indirectly assessed by remote sensing with a grid size of 3 km by 3 km. Triggers below which payments are made correspond to percentile 5% of historical long running decadal relative evapotranspiration data. Threshold for full payment is adjusted depending of areas and crop development period. Yet pay-outs are dependent on three specific periods mimicking the different stages of maize production (since 2012). The first stage covers 30 days after seeding (1st of July), the second stage comprises 20 days and the last stage 40 days. Pay-outs are proportionally to the total covered amount for the three subsequent stages are 30%, 100% and 100% respectively.

The insurance case in Burkina Faso focuses on an index-based insurance scheme issued by PlaNet Guarantee. The scheme covers drought risks in maize. Maize is selected since it requires relative high amounts of inputs and output is more volatile than for example millet and sorghum which are more resistant to drought. The system works by a combination of crop insurance and a rural credit facility. The credit agency insures their portfolio of loans whereby the lenders sign in addition to the loan contract an accompany insurance contract. The pay-outs are made via the credit agency but is withhold if the credit is not returned.

In Burkina Faso the scheme was launched by a pilot with 194 maize producers during the 2011 season by PlaNet Guarantee while in 2013/2014 2,072 producers were insured. Producers pay a premium of 9.5% of the loan amount requested since 2012. The premium is not differentiated between covered zones and regions, but each zone and region has its specific threshold level (and thus actuary fair). This implies that protection levels are higher in the South which is less drought risk prone. The micro finance institution also requests a deposit amount of 25%.

Table 2: Key characteristics of maize index-based insurance scheme in Burkina Faso.

Year 1/	Number of farmers	Hectares	Premium	Premium (%)
2011	194	227	35,000	10.0
2012	1,340	1,507	155,000	10.8
2013	1,885	1,813	145,000	10.8
2014	2,072	2,212	210,000	9.5

1 Subscription around July, harvest around October, pay-offs if any before the end of the year.

PlaNet Guarantee seeks to extend the experiment conducted among 10,000 producers for subsequent seasons, expanding to other crops. In 2013 the PlaNet Guarantee insurance scheme was extended to cotton production, and 446 producers took up the product. Note that the Burkina Faso project is part of a larger project whose objective is to develop parametric agricultural insurance systems in WAEMU countries, including Senegal, Mali and Burkina Faso. This facility should cover at least 60,000 people by the end of 2015 in West Africa and raise awareness to more than 165,000 farmers on agricultural insurance.

Source: Van Asseldonk et. al. 2015. Evidence of linking crop insurance and rural credit by FARMAF Policy Brief No. 1 p. 4. http://www.farmaf.org/images/documents/Operational_materials/policy_brief/FARMAF_Policy_Brief_Insurance_2015.pdf

6. Assessment Test for Tool # 5: Weather index based insurance

A farmer who has heard about a weather index insurance (WII) scheme being promoted by the government in his area approaches his District Extension Officer (DEO) for advice. The DEO explains to the farmer whether each of the following descriptions of the WII scheme is appropriate (accurate) or not.

1. WII is suitable to transfer crop and livestock production risks arising mainly from abnormal weather such as droughts, floods or detrimental levels of temperatures, but not from other factors such as hail, earthquakes, fires, etc.

2. The most important advantage of WII is the relatively low operational and transactions costs since there is no need for a costly on-farm loss assessment.

3. The WII insurance payouts can be much quicker given that the measurement of weather station data, with no adjustment of individual farm production performance is required.

4. Index insurance requires that all insured farmers within the defined area have the same insurance payout conditions, regardless of their specific risk exposure, thus there is a higher chance of "adverse selection".

5. There is a lower chance of "basis risk" in WII. A basis risk is the difference between the loss experienced by the farmer and the payout triggered, meaning the chance that the payment was triggered but no actual losses occurred or that the index was not triggered but losses occurred.

Answers

1. Accurate. The peril has to be easily measurable, highly correlated to production performance. WII can be adopted for both crop and livestock production with the selection of appropriate indices.
2. Accurate. Estimation of an individual farmer's field loss is not required.
3. Accurate, when reliable and timely weather data is available.
4. False. WII leads to a lower chance of adverse selection, not higher.
5. False. Higher basis risk in WII.

7. References:

1. **Chantarat, S., Barrett, C. B., Mude, A. G., & Turvey, C. G.** 2007. Using weather index insurance to improve drought response for famine prevention. *American Journal of Agricultural Economics, 89*(5), 1262-1268.

2. **FAO**. 2013. Rainfall insurance: "weather" it rains or not
http://www.fao.org/gender/gender-home/gender-why/bite-sized-stories/rainfall-insurance-weather-it-rains-or-not/en/

3. **FAO.** 2006, Mar. *Development of a Weather Yield Index (WYX) for Maize Crop Insurance in Malawi,* by Chavula, Adams & Gommes, René. Working Paper.
http://www.fao.org/nr/climpag/pub/Maize%20crop%20Insurance%20in%20Malawi%202006%20Chavula%20Gommes.pdf

4. **IFAD.** 2011. *Weather Insurance for Farmers: Experience from Ethiopia*, by Araya, Nahu Senaye. Rome, Italy.
http://www.ifad.org/events/agriculture/doc/papers/araya.pdf

5. **IFAD** and **WFP.** 2011. *Weather Index-based Insurance in Agricultural Development: A Technical Guide.* Rome, Italy.
http://www.ifad.org/ruralfinance/pub/wii_tech_guide.pdf

6. **IFAD and WFP.** 2010. *The Potential for Scale and Sustainability in Weather Index Insurance for Agriculture and Rural Livelihoods*, by P. Hazell, J. Anderson, N. Balzer, A. Hastrup Clemmensen, U. Hess and F. Rispoli. Rome, Italy.
http://www.ifad.org/ruralfinance/pub/weather.pdf

7. **Kerer, Jan**. 2013. Background Paper on the Situation of Agricultural Insurance in Kenya with Reference to International Best Practices. *Adaptation to Climate Change and Insurance (ACCI)* Nairobi, Kenya.

8. **Luxbacher, Kirsten and Andrew Goodland**. 2012. "World Resources Report Case Study. Building Resilience to Extreme Weather: Index-Based Livestock Insurance in Mongolia." World Resources Report, Washington DC.
http://www.worldresourcesreport.org

9. **Swain, Mamata.** Dec. 2015. Performance of Crop Yield and Rainfall Insurance Schemes in Odisha: Some Empirical Findings. Agricultural Economics Research Review. Vol. 28 (2); pp 201-211.

10. **Van Asseldonk, M., Porgo, I., Hamusimbi, C., Mumba, G., Banda, J., Maitre d'Hotel, E., le Cotty, T., Burger, K., and Onumah, G.** 2015. *Is there evidence of linking crop insurance and rural credit and its potential benefits?* Policy Brief No 1, 2015. The Farm Risk Management for Africa (FARMAF) Project.
http://www.farmaf.org/images/documents/Operational_materials/policy_brief/FARMAF_Policy_Brief_Insurance_2015.pdf

11. **World Bank/International Finance Corporation.** 2015. Kilimo Salama - *Index-based Agriculture Insurance : A Product Design Case Study.* Washington, DC.
https://openknowledge.worldbank.org/bitstream/handle/10986/21682/947940WP0Box380ture0Insurance0Final.pdf?sequence=1&isAllowed=y

12. **World Bank.** 2012. Weather Index-based Crop Insurance in Malawi: Facilitating Farmers' Access to Agricultural Credit, by Sadler, Marc & Mahual, Oliver.
http://siteresources.worldbank.org/EXTDISASTER/Resources/Malawi_WeatherInsurance_Final.pdf

13. **World Bank.** 2011a. Weather index insurance for agriculture: Guidance for development practitioners. Agriculture and Rural Development Discussion Paper 50, Washington D.C.

https://www.agriskmanagementforum.org/sites/agriskmanagementforum.org/files/WEATHER%20INDEX%20INSURANCE%20FOR%20AGRICULTURE_WB_WII_Paper_Nov_2011.pdf

14. **World Bank.** 2011b. *Index-based livestock insurance in Mongolia: Protecting herders from Climate-related livestock mortality*, by Goodland, Andrew & Mahul, Olivier. https://www.gfdrr.org/sites/gfdrr.org/files/DRFI_Mongolia%20IBLIP_Final.pdf

15. **World Bank.** 2007. Designing Weather Insurance Contracts for Farmers In Malawi, Tanzania and Kenya. Columbia University. http://iri.columbia.edu/~deo/IRI-CRMG-Africa-Insurance-Report-6-2007/IRI-CRMG-Kenya-Tanzania-Malawi-Insurance-Report-6-2007.pdf

Tool # 6: Agricultural Finance and Microfinance

1. What is it? – Concepts and Characteristics

Financial markets for agriculture in most developing countries are constrained by two main factors. Formal credit offered through official banks and other credit institutions is not well established, not accessible in rural areas and requires collateral in the form of real or liquid assets which most small, subsistent and tenant farmers may not have. Secondly, the informal credit from private money lenders usually carries an exorbitant interest, making it unaffordable or not profitable. Thus, due to a lack of access to credit, farmers may not be able to invest in improved seeds, fertilizer and other technologies that are considered necessary to enhance farm productivity and move households out of the poverty trap. Therefore, having access to secure and affordable credit for farming is potentially an important tool for coping with risks in agriculture. The main objective of this tool is to understand how access to credit for agricultural activities is likely to help in minimizing and managing different types of risks in the sector.

Undoubtedly, financial services play a critical role in reducing poverty[8] and managing risks, including basic services such as secure savings accounts. Permanent access to financial services can help poor people take control of their lives. Good management of very small assets can be crucial to the survival of poor people who live in precarious conditions, threatened by a lack of income, shelter and food. To come out of poverty, farmers need to be able to borrow, save, invest and protect their family income against risk. Thus, access to secure credit can be a risk management tool for agricultural enterprises.

Banking and other financial services are not easily accessible to farmers, entrepreneurs and small businesses in remote rural areas or low income clients throughout the country. Microfinance Institutions (MFIs) were primarily created to provide credit and other financial services to such clients. They offer basic financial services (typically loans and savings but some may offer a wide range of individualized financial services) to individual clients or a group of clients as one unit.

Since the success of "Grameen Bank" in Bangladesh during the early 1980s, MFIs have become popular in many developing countries. Some have evolved from non-government organisations (NGOs) working in poor communities with a strong social service orientation. Their transaction costs vary a great deal and therefore their charges also range from very low to very high, especially given that they deal with very small amounts per account and whether they themselves receive external aid/development funding.

MFIs can provide services in a cost efficient way, especially if they are able to provide loans without collateral and reduce the transaction costs of dealing with accounts involving meagre sums of money. MFIs that follow the group loan approach (such as Grameen Bank, Village/Community Bank, informal savings groups, etc.) use peer pressure of the members in the group to minimize problems of moral hazard, adverse selection and loan defaults (see for example, IFAD 2005 project in Gambia which supported more than 30 village bank networks). In general, membership-based organizations can facilitate access to financial services and be viable in remote rural areas (IFAD 2005c).

A **rotating savings and credit associations**, or **ROSCAs**, are community level informal or 'pre-co-operative' microfinance groups that have been in operation in most developing countries. It involves a group of individuals who agree to meet for a defined period in order to save and borrow

[8] See IFAD (2008).

together and thus perform a financial service of credit provision in an informal way. The success of this collaborative finance model has been based on the peer pressure and peer involvement.

The use of mobile communication technology for the delivery of financial services in developing countries has been increasing, enabling a greater outreach to the rural population. FAO has piloted a mobile financial services scheme in Zimbabwe since 2007 and Stewart Bank and Econet Wireless have been offering savings and loans services since 2011 to over a third of the rural subscribers in Zimbabwe. This innovative business model of reaching a large number of users at relatively low transaction costs by using existing mobile phone networks and involving other stakeholders in the value chain such as input suppliers, is very promising for its potential in less developed and developing countries.

A report by Deutsche Bank (2007) emphasized that the evolution of microfinance into a niche investment product attracts both retail and institutional investors. Microfinance is also said to benefit from a further strong rise in socially responsible investments. In 2007, Forbes Magazine published a list of the 50 Top MFIs in the world out of the 641 reviewed[9]. Based on the scale, efficiency, risk and returns criteria, Bangladesh and India had 7 each in the top 50, followed by Bosnia and Herzegovina, Morocco and Peru with four each and Colombia with three. Among African countries only Egypt, Ethiopia and Morocco were listed as having MFI(s) in the top 50. Of course there will be many more in the full list (unfortunately, it is not published).

As stated by IFAD (2009), "Innovations in financial services, particularly in microfinance, have enabled millions of women and men in rural areas – formerly excluded from the financial sector – to gain access to these services on an ongoing basis." The Grameen Bank for example, serves mostly women, 97% of their over 8 million borrowers.

The access to much needed financial services, including agricultural credit, can be enhanced very efficiently with the public-private partnership in designing and implementing programs under this tool.

2. Suitability of the tool

Agricultural finance is suitable as a risk management tool only for those farmers who can make productive use of the funds borrowed. In general, cash crop farmers or market oriented small or large farmers are more likely to use this tool. However, the tool could be just as beneficial to smallholder farmers and food crop producers provided they follow the basic financial management of their activities.

Before making the decision to access credit, the individual farmer or a small business entrepreneur needs to consult with different advisors, namely agricultural extension agents, farm management specialists, local financial institution staff or other experts. In order to use the services of agricultural finance and microfinance institutions for effective agricultural risk management, farmers need to have a credible business plan before they embark upon obtaining finances. It is vital for the entrepreneur to find out answers to questions such as:

- Is acquiring loan the best option, is it going to add to your profitability and what are the expected benefits, costs and potential risks?
- Who are the providers of credit and financial services formal and semi-formal institutions in the region? What is their trustworthiness and what are their requirements (terms and conditions, fees, charges, collateral requirement, repayment flexibility etc.)?

---

[9] Source: http://www.forbes.com/2007/12/20/microfinance-philanthropy-credit-biz-cz_ms_1220microfinance_table.html

- Do they offer other financial services such as savings/chequings accounts, lines of credit, fund transfers (including subsidy or other safety net payments), the nature of collateral requirements, insurance, etc?
- Are there any special programs offered and if you are eligible, any aid or development projects in the area and how can you take advantage of them?

3. Advantages

Access to banks and other formal institutions to receive agricultural finance services can be very useful for farmers in realizing truly productivity/profitability enhancing projects. Financial services can potentially be used to protect from most sources of risk. MFIs may provide services that are otherwise not available elsewhere, even if they are run by NGOs or non-profit organizations.

Microcredit loans can also obtain relatively high rates of return mainly due to the scale effect. For example, if a small street vendor buys his commodity for 50 cents and sells it for a dollar, he gets a return of 100%. However, generally speaking, this is not possible for a large business.

Microfinance is also seen as a tool to help reduce income inequality and help reduce poverty by focusing on the poor and helping them to participate in the economy (Hermes).

4. Disadvantages

Although most agree that there is an overall positive effect of microfinance such as school enrolment, women empowerment, and making people's lives better (Banerjee et al., 2015), it has also been criticised by many (e.g. Cooper 2015) for all the problems it faces including loan defaults, low net returns on small loans due to high transaction costs (on average over 30% interest on loans and negative returns on savings) and not having a significant impact on poverty reduction overall.

As stated by Skees (2003), a major advantage of microfinance institutions has been the ability to pool risk as long as the risk is not correlated. However, risk in agriculture can be highly correlated, meaning a large number of households are likely to face bad outcomes (poor yields, unfavourable prices, etc.) together. That's why small rural finance entities are not able to pool and manage correlated risk on their own. On a more positive note, two innovations offering the most hope are: 1) the use of global futures markets by intermediaries who can offer a form of price insurance; and 2) the use of index insurance contracts to shift natural disaster risk into the global markets. Hence, blending these forms of index insurance and rural finance is seen as the ideal strategy.

There can be problems of trust with MFIs and NGOs arising mainly from poor regulations, political interference, high inefficiencies, poor institutional infrastructure, lack of technical, managerial expertise, etc.

However, many of these problems are solvable by relying on innovations through public and private mechanisms to make credit and financial services available as efficiently as possible while tying these to some forms of national/international public support.

5. Case Studies

CASE STUDY 6: Equity building society of Kenya reaches rural markets (IFAD)

This case study outlines the strategy employed by Equity Building Society in Nairobi and Kenya's Central Province to expand its operations for low and moderate-income farmers in remote rural areas. It now provides loans to medium- and small-scale tea and dairy farming, secured by the farmer's contracts with agribusinesses. By developing a more customer and market driven approach, with assistance from international donor organizations, Equity Building Society managed to prevent the bank's collapse by providing mobile banking and adopting a microfinance model in agriculture. This is one of the key standard ARM tools and when available at reasonable terms and conditions, can modernize farming operations and help farmers to manage various risks in agriculture.

Equity Building Society of Kenya reaches rural markets

Summary

Equity Building Society provides microfinance services to more than 250 000 low- and moderate-income citizens in Nairobi and Kenya's Central Province via a network of branch offices and mobile banking units. After a decade of extending long-term mortgage loans to an untargeted clientele with meagre results, Equity altered its approach. In 1994, the bank began tailoring its loan and savings products to a microfinance market, eventually adding two loan products for tea and dairy farmers. The products are secured by agribusiness contracts.

By the end of 2003, Equity's deposit base had grown to USD 44 million, and loans topped USD 21 million. Thanks in part to strong donor support, Equity instituted a mobile banking programme in 2000 with the purpose of efficiently reaching more clients in remote rural areas. Mobile operations have been successful, covering costs and accounting for more than USD 1.3 million in deposits in 2003. Mobile banking has also decreased branch congestion and improved the brand image and visibility of Equity.1 This case study outlines the strategy employed by Equity[1] to expand its rural outreach.

Background

Meeting the challenges of rural services

In order to expand its client outreach sustainably in rural areas, Equity developed demand-driven products for farmers, improved operational efficiency and introduced mobile banking operations. The groundwork was laid for this through a much-improved management information system.

Developing products that responded to rural customer needs

Two new products were specifically developed to meet the needs of agricultural customers: a crop advance loan and a farm input loan. These loans range in size from USD 21 to USD 820 and are made to tea, coffee and dairy farmers for terms of up to one year. The crop advance loan is based on expected farm production and is intended to smooth consumption or permit farmers to invest in another enterprise before crop sales have been realized. The farm input loan is intended for the purchase of fertilizer, pesticide, farm labour and other seasonal inputs. Farmers pay commercial interest rates of 21 to 24% annually on these loans, charged on the declining loan balance.

To mitigate risk, both types of loans must be secured by agribusiness contracts and are made only to farmers receiving regular payments by produce processing and marketing companies. Kenya Tea Development Agency, for example, processes, markets and sells tea on behalf of farmers, paying them a monthly advance over the course of the growing season, followed by an annual bonus based on the actual proceeds of the processed tea sales. In order to be eligible for Equity loans, farmers open a savings account with Equity, to which Kenya Tea Development Agency transfers monthly and bonus payments. Upon approval, Equity loan proceeds and automatic loan repayments are posted to this account, backed by the contractual payments of Kenya Tea Development Agency.

Reaching clients in remote areas

As Equity regained its footing in the late 1990s through the provision of more demand-driven products, it began to focus on building a larger clientele. Establishing permanent branches in remote rural areas was not financially tenable, so Equity struck on the idea of mobile banking units. The effort began with USD 262 000 of Equity's own capital in 2000. Two four-wheel drive vehicles were put into service in rural areas with population densities of over 400 people per k^2. As the viability of the mobile units became clear, a USD 411 000 grant from the Department for International Development (UK) permitted Equity to integrate more sophisticated technology and increase the scale of its mobile banking outreach.

The mobile units make weekly visits to each community that they serve and provide most services available at Equity branches, including processing and administering five different savings products and several loan products, as well as general financial services (such as the sale of cashier's checks and interbranch cash transfers). Mobile customers pay a modest monthly fee for the service (lower than bus fare to the nearest branch) to help Equity cover related expenses.

As of early 2004, Equity's mobile units profitably served 29 rural villages and about 12 000 clients on a regular basis.[5] A number of special features enable these units to operate successfully in the rural Kenyan environment:

Donors and investors

Equity Building Society is a privately held Kenyan company. The majority of its equity (84%) is held by Equity management, staff and clients. In 2003, the international equity investment fund AfriCap purchased 16% of the company. Equity proactively sought international assistance to accomplish its goals, and donors responded to the institution's focused requests. Since 1993, the bank has received donor and investment assistance in the following forms:

One of the most important factors in donor funding has been a commitment to provide Equity leadership and staff with the skills and expertise necessary to make sound, strategic decisions on their own.[14] Donors have focused on building institutional capacity, while standing back to allow Equity to develop its own products and implement necessary changes.

Lessons learned

The following lessons can be gleaned from Equity's experience in reorienting its services towards microfinance clients, including rural borrowers and savers.

- Developing a market-driven, customer-focused approach was a prime factor in turning around a poorly performing portfolio, increasing deposits and reaching rural customers.
- Mobile banking units can be a viable answer to the problem of reaching isolated clients in rural areas.
- Flexible rural delivery mechanisms such as mobile banking need not be restricted to credit and deposit services, but can be used to offer a number of different services, including inter-branch cash transfers.
- Loans secured by contractual agribusiness payments can be an effective method of extending credit to farmers without taking on extensive risk, although this practice can exclude many smaller farmers.
- A sound management information system provides the basis for any product or technological innovation. The combination of a reliable management information system and innovation can improve institutional efficiency, lower the cost of service delivery, increase customer satisfaction and perception and allow for greater product customization.
- The initial expense of new technology, such as a management information system and mobile banking units, may justify initial donor grant support.
- Emphasizing institutional capacity-building is a proven donor tactic for effectively supporting long-term MFI sustainability and growth.

Conclusion

The successful turnaround of Equity between 1993 and 2003 shows the tremendous institutional adaptability and commitment of the MFI. Its mobile banking operation provides rural Kenyan citizens with access to secure financial services at an affordable cost and on a consistent basis and has increased both the deposit base and the profitability of the bank. Many former branch customers are now opting to use mobile services, resulting in reduced congestion and improved efficiency at Equity branches, as well as better economies of scale among the mobile units.[15] Although the mobile units are reaching an increasing number of rural borrowers, primarily farmers who save and borrow for consumption, agriculture and other business purposes, Equity currently offers no loan products designed for small independent farmers without agribusiness contracts. This represents a market opportunity, as over half of Equity's clients depend on agricultural income, mostly through medium- and small-scale tea and dairy farming.[16]

Equity is currently attempting to reach further down-market by tailoring new services to poorer clients and smaller entrepreneurs. Loan sizes currently range from just USD 6 to over USD 30 000, and effective annual interest rates vary from 21 to 53%. This trend towards poorer clients can be seen in increasingly smaller loan sizes and a new savings account with no minimum balance, introduced in 2004, for small farmers. The new account, offered in partnership with the NGO Pride Africa, will allow lower income farmers to access Equity services for the first time.

Source: Excerpts taken from **IFAD. 2005**. Emerging Lessons in Agricultural Microfinance: Selected Case Studies. pp. 44-50, by Douglas Pearce and Myka Reinsch.
http://www.ifad.org/ruralfinance/pub/case_studies.pdf

6. Assessment Test for Tool # 6: Agricultural finance and Microfinance

Consider the following farming business situations that a given farm family or a group of farmers are faced with. In each case, please indicate if acquiring formal credit (getting a loan) would reduce or increase the risk for the borrower (farmers). The answer could be yes (most likely would reduce the risk, so it is a good risk management decision), no (most likely would increase risk) or maybe (it would still be a good risk management decision, provided certain necessary steps are taken).

1. A farmer wishes to dig a deep well (or tube well) on his farm as some of his neighbours have done. This will help to irrigate the fields and produce multiple crops. However, it would require a relatively large investment loan and the water availability (or the depth at which water source is accessed) cannot be guaranteed.

2. A farmer wishes to get an operating loan for purchasing seeds, fertilizer and other important inputs to carry out farm production activities.

3. A microfinance loan (with a business plan prepared with the help of a local NGO) for a group of farm women on a small scale, for example, to buy a goat or a cow, make cottage cheese and other processed milk products.

4. The formation of an Agricultural Marketing Cooperative to sell farm produce in local/area markets with collective bargaining.

Answers

1. No. This is a relatively risky investment and would increase the risk for the farmer. However, this is a "higher-risk-higher-expected income" situation. Thus, the final risk management decision would depend on the risk preference of the farmer.
2. Yes
3. Yes
4. Maybe. Normally this would reduce the risk by getting higher prices and increasing certainty. However, a successful organization (with a credible business plan) and functioning (with management) of the Cooperative are essential but difficult elements.

7. References:

1. **Banerjee, Abhijit, Karlan, Dean and Zinman, Jonathan.** "Six Randomized Evaluations of Microcredit: Introduction and Further Steps." *American Economic Journal: Applied Economics* 7.1 (2015): 1-21. Print.

2. **Cole, S., Giné, X., Tobacman, J., Topalova, P., Townsend, R., and Vickery, J.** 2012. Barriers to Household Risk Management: Evidence from India. IMF Working Paper, Research Department.

3. **Cooper, Logan,** 2015. Small Loans, Big Promises, Unknown Impact: An Examination of Microfinance. The Apollonian Revolt journal. http://theapollonianrevolt.com/

4. **Deutsche Bank,** Dec. 2007. Microfinance: An emerging investment opportunity. http://www.newswire.ca/news-releases/deutsche-bank-announces-key-findings-in-microfinance-study-534916521.html

5. **FAO.** 2015. Mobile technology applications in agricultural finance: global trends and innovations in Sub-Saharan Africa. **Webinar presentation** by Emilio Hernandez, 11 Dec. 15 2015.

6. **FAO.** 2013. *Four case studies on credit guarantee funds for agriculture.* Rome.

7. **Hermes, N.** 2014. Does microfinance affect income inequality? Applied Economics, 46(9), 1021-1034.

8. **IFAD.** 2015. Sept. *Rural Finance: Small amounts making a big difference.* http://www.ifad.org/events/microcredit/eng.pdf

9. **IFAD.** 2010, Mar. Decision tools for rural finance. Rome http://www.ifad.org/ruralfinance/dt/full/dt_e_web.pdf

10. **IFAD.** 2009, Aug. *Gender and Rural Microfinance: Reaching and empowering women. Guide for Practioners ,* by Mayoux, Linda & Harti, Maria. http://www.ifad.org/gender/pub/gender_finance.pdf

11. **IFAD.** 2008. Microfinance: A lifeline for poor rural people. *Article by Lennart Båge, President of IFAD published in Uluslararası Ekonomik Sorunlar, a bi-monthly journal of the Turkish Ministry of Foreign Affairs.*

12. **IFAD.** 2005a. Banking on Villagers - The Gambia: Rural Finance and Community Initiatives Project. http://www.ifad.org/evaluation/public_html/eksyst/doc/profile/pa/gm.htm

13. **IFAD.** 2005b. *Emerging Lessons in Agricultural Microfinance: Selected Case Studies.* http://www.ifad.org/ruralfinance/pub/case_studies.pdf

14. **IFAD.** 2005c, Aug. *Managing risks and designing products for agricultural microfinance,* by Christen, R. P., & Pearce, D. Consultative group to assist the poorest (CGAP). http://www.ifad.org/ruralfinance/pub/risks.pdf

15. **Skees, J. R.** 2003, Jun. *Risk management challenges in rural financial markets: blending risk management innovations with rural finance.* In the thematic papers presented at the USAID Conference: Paving the Way Forward for Rural Finance: An International Conference on Best Practices, Washington, DC (pp. 2-4).

LESSON 3
Market Related Risk Management Tools

Learning objectives

At the end of this lesson, learners should be familiar with the concepts involved in various market and marketing related risk management instruments, appreciate the opportunities/advantages/benefits and challenges/disadvantages/constraints of the selected tools and judge their suitability in a given area or farm situation.

In today's agriculture, the marketing of farm commodities plays a very important role in the financial success and wellbeing of a farm family. A farm may be highly productive, but if the farmer is not able to market the produce as efficiently as possible by getting remunerative prices and minimizing/mitigating possible market risks, it will not achieve financial success in the long run. Tools outlined in this section, therefore, deal with the various means of getting the best possible price and secured access to the market with the minimum possible risk.

These market related instruments are grouped into three broad categories as follows:

Tool 7: Contract Farming
Tool 8: Commodity Exchanges and Futures Markets
Tool 9: Warehouse Receipts Systems

Contract farming has grown significantly around the world and some see it as a key instrument in modernizing the agriculture industry, achieving socio-economic development and reducing poverty in developing countries. It has also been criticised where it has failed to achieve its objectives and has been a cause of exploitation. The generic term "contract farming" is used to cover marketing contracts, organized group marketing and different types of contractual arrangements between the primary producer (generally a framer) and a firm (generally a processor, large warehouse chain) interested in procuring farm production as a raw material.

The second tool provides a profile of the main functions of a commodity exchange and focuses on the futures contracts useful for hedging against price volatility faced by farmers. It intends to give a basic understanding of the way commodity exchanges and future markets work and how they can help manage price risk. The module explains how this tool works and what it can do. Although the marketing of agricultural commodities through formal, organized commodity exchanges is growing around the world, it is still not very common in Africa. Currently, there are only three operational agricultural commodity exchanges in Africa (South Africa, Malawi and Ethiopia). However, combined with futures contracts facilities, they can provide very sophisticated and innovative options for farmers. This is not a risk management tool for all farmers but it is useful to understand their potential where such facilities are available and other places where it could potentially be developed and sustained.

In the African context, the tool on warehouse receipts system (WRS) is primarily about securely storing crops both physically and economically. It becomes a marketing tool when the primary goal is to sell the produce at the most opportune time in the best way possible. It also serves a dual function of securing financing from a credit institution by using the receipts as collateral. This tool explains the role of warehouses as instruments to manage seasonal price risk. It provides an explanation as to how warehouse receipts systems can help with collateral to access credit.

Not all market related tools will be applicable to every farm or locality but there may be some that are suitable and accessible in a specific situation. For other stakeholders, including policy makers, an understanding of all available modern tools is strongly recommended.

Finally, similar to other lessons, this lesson does not cover all possible tools. For example, practices of direct marketing of farm products, on line marketing, marketing through fair trade, institutional arrangements for storage at farm, community, cooperative or district levels, etc. are not covered explicitly.

Key words for lesson 3:

1. *collateral manager,*
2. *commercial stakeholders,*
3. *commodity exchange,*
4. *community inventory credit,*
5. *contract farming,*
6. *futures contract,*
7. *futures option,*
8. *hedging,*
9. *market transparency,*
10. *out-grower schemes,*
11. *processors,*
12. *smallholder farmers,*
13. *storage*
14. *value chain*
15. *warehouse receipt*

Tool # 7: Contract Farming

1. What is it? – Concepts and Characteristics

Risk related to certain aspects of farming (production and marketing) can be shared by farmers with other firms, under contractual arrangements. These arrangements are sometimes called contract farming, marketing contracts, and collective marketing arrangements among others. The system of agreements, arrangements and contracts that improves links between farmers and consumers, typically through one or more intermediaries, has the potential to reduce price and marketing risks for small farmers.

Hence, the objective of this tool is to explain how various types of contract farming schemes can be used as an instrument of agricultural risk management along with their pros and cons and applicability in a specific given farm situation.

There are several types of contractual arrangements between small-scale producers and commercial stakeholders. They may involve traditional contract farming, large multinational supermarket chains contracting with a wide range of farmers, out-grower schemes, fair-trade schemes or multi-stakeholder coordination along the value chain.

Under a broad economic approach, **contract farming** generally refers to "a particular form of supply chain governance adopted by firms to secure access to agricultural products, raw materials and supplies meeting desired quality, quantity, location and timing specifications." (UNIDROIT/FAO/IFAD, 2015).

USDA's Risk Management Agency (RMA) defines **a marketing contract** as a contract between a farmer and a processor or handler that establishes a marketing outlet and formula for determining the price of a commodity before harvest or before the commodity is ready to be marketed. Similarly, a **production contract** is defined as an agreement between the farmer and a processor. It usually details the production inputs supplied by the farmer and the processor, the quality and quantity of a particular commodity that is to be delivered, and compensation that the farmer will be paid. In return for giving up control over decision making, the farmer is often compensated with a price premium or lower market risk.

The concept of contract farming, which often involves a marketing contract, can be very broad as agreements can vary widely. Under the traditional form of contract farming, farmers produce their crop and deliver it to a buyer (usually a miller, processor, warehouse/grocery chain etc.) as per the agreed quality, quantity and delivery timetable. Usually the purchaser, in turn, agrees to accept these goods at a predetermined price. In many cases, the provision of inputs, credit, technical know-how and other services may be stipulated to be provided by the purchaser. Thus, contract farming, as a general term, may include a variety of arrangements between a farmer and a firm, ranging from just the procurement of the farm produce to a full service-full control of farming operations.

An **Out-grower scheme**, one of the more modern practices of contract farming, involves a closer link of many small farm operations with a processor or a large farm/firm which supports or controls production planning, input supply, technical advice and transport. There may be a greater uniformity of standard of the produce produced under this scheme. This type of arrangement provides guaranteed market access and covers some or all non-labour costs for the producer.

The primary purpose of contract farming is to reduce or eliminate the price and market access risk for the farmer. Contract farming or marketing contracts transfer risks from farmers to the other contracting party. This reduces price/marketing and financial risks, especially if the provision of necessary operating and investment capital is involved, and production risk if the supply of high quality/improved seeds, fertilizer, chemicals, etc. is involved. It may tie up with insurance against agricultural production hazards.

Contract farming can be a very useful and important tool for farmers to manage agricultural risk. However, the benefits of the specific contract depend on the details of the contract, the existence of loopholes, how well they are implemented and the overall legal/regulatory framework to govern the regime. To be sustainable, contractual marketing arrangements need to benefit both parties.

Another marketing tool, similar or closely linked to contract farming, which is useful in managing agricultural risk, is **collective marketing,** where small-scale producers form groups, associations or cooperatives for the purpose of marketing their common product(s). This helps them to manage price/marketing risk by enabling their access to domestic, regional and international markets, reducing their transaction costs and improving their bargaining power (IFAD/FAO 2012). Through the improved bargaining power they may be able to formulate a more advantageous marketing contract with the global supermarket, national chain or processor better than an individual small farmer.

There is a need and potential for public private partnership in designing and implementing programs under this tool.

2. Suitability of the tool

Contract farming or marketing contracts are generally suitable for cash commodities such as vegetables, fruits, cotton, tobacco, sugarcane, dairy and poultry. They are less practiced for staples food crops such as cereals, root crops and pulses. Many products desired by large supermarket chains are ideal for marketing contracts. The out-grower schemes, on the other hand, are more suited for plantation crops, seed production, etc.

Because of these peculiarities, this tool is more suitable for market orientated farmers, small or large, rather than subsistence food growers. However, evidence from Southeast Asia suggests that small scale paddy producers also engage in contract farming. The literature (Minten et al 2009) examining evidence from Madagascar suggests that given the right incentives and contracting systems, small farmers in developing countries, from Africa in particular, can participate successfully in these emerging value chains.

This tool is more suitable for managing market risk, the price and secure access to the market. However, the contracts that involve input deliveries and technical advice, can also help farmers reduce production risk.

In view of the ability of contract farming arrangements (CFA), especially with the synergetic effect in the presence of the collective marketing organization, national and international policy makers find this tool to be an attractive option to help poor farmers in developing countries (Barrett et al. 2012). Similarly, vigilant governments also recognize the potential opportunities of modern contractual arrangements in agriculture to develop this sector and provide a policy enabling and regulatory environment in which to promote them.

3. Advantages

An OECD study by Vavra (2009) concluded that "contract farming appears to be the main road towards making African agriculture more market-oriented".

In addition to the obvious potential benefits of price certainty (or price discovery transparency) and assured market access, according to UNIDROIT/FAO/IFAD (2015) and others, contract farming is generally recognised for its potential to sustain and develop the production sector by contributing to:

- capital formation,
- introducing and transferring new technology and skills
- input supplies
- access to credit, technical know-how and advisory services,
- increasing agricultural production and yields,
- economic and social development,
- environmental sustainability,
- increased reliability in supply quantity and quality and,
- improving the traceability of the products, which is increasingly important from a food safety point of view.

Small farmers benefit because of a combination of effects, such as improved access to inputs, credit, extension services, technology adoption, and from productivity spill over effects on other crops and enhanced income stability (Minten et al 2009).

The final consumers and other supply chain participants may also benefit from successful contracting arrangements as these may lead to a varied and stable supply of raw material as well as the farm produce. Consumers may also experience improved quality and better prices for household food commodities.

4. Disadvantages

Although there are many benefits from contractual arrangements in agriculture, many small farmers in developing countries may experience the adverse effects of such contracts. Cheating and breach of contract on the part of either party, leading to mistrust and failure of the system is the biggest challenge. Those with less bargaining power, generally unorganised small holder farmers, are the vulnerable party in this type of risk management strategy.

A study by ActionAid (2015) summarised that "Out-grower schemes are an important component of many current public-private partnerships in developing countries. Such schemes often appeal to farmers because the company often provides inputs and production services. In addition, farming incomes can rise and such schemes often open up new markets and provide new technology." However, their review also indicated that "Criticisms of out-grower schemes and contract farming reveal the unequal power relationship between a company and farmers. Farmers often provide both the land and cheap labour, and at the same time carry most of the risk."

UK's DFID's criticism of contract farming schemes mentions that they "often exclude the poorest farmers including the landless, marginal and subsistence farmers and women".

According to the study by UNIDROIT/FAO/IFAD (2015), typical complaints against contract farming can be the following:

- Small farmers do not find the arrangement financially advantageous, (they may find themselves on the losing side of the deal)
- Generally, contract enforcement and dispute resolutions are difficult, very lengthy, costly, and thus often avoided by parties.
- Improper use of the credit provided by the contractor might lead to unsustainable levels of indebtedness for the producer.
- Changes in working conditions may affect the producer's family or workers, and this raises concerns in certain countries.
- If not adequately protected, women may not benefit fully from the potential advantages that could accrue from contracts.
- Switching from subsistence farming to cash crops might also cause problems related to monoculture production such as the loss of biodiversity and even a threat to the producer's own food security.

5. Examples/Case studies

CASE STUDY 7: Egypt: Smallholder contract farming high-value & organic exports (IFAD)

This case study reports on the potential for involving smallholders in the production of non-traditional high-value horticultural crops for export within the context of contract farming in Egypt. The development strategy focuses on the inclusion of smallholders in the horticultural export value chain and fruit and vegetable processing through contract farming arrangements. The stakeholders of this project are farmers, farmer associations, exporters, processors, the Government and the donors (the German Agency for Technical Cooperation, USAID, and IFAD). The case study reveals that smallholder families could see substantial increases in their incomes by engaging in the contract farming of organic horticultural produce as well as in conventional export crops, thus enhancing the resilience of farmers and their ability face/manage various risks in farming.

Egypt: Smallholder contract farming for high-value and organic agricultural exports

Executive summary

Background

The total population of Egypt has grown rapidly over the last 25 years, to more than 70 million, but the annual rate of population growth has slowed to a moderate 2.1 per cent. As a result of structural reforms in the 1990s, the Egyptian economy experienced a period of sustained growth between 1994 and 2006, with a short spell of recession in 2003; however, this was reversed after the arrival of the new reformist government in the summer of 2004. The economy has improved considerably since.

Trade in high-value products, such as fruits and vegetables, is increasingly displacing exports of traditional commodities. Thus, during the 1990s, the aggregate value of world trade in traditional agricultural commodities declined, while the value of trade in high-value exports grew by around 6.8 per cent annually (approximately 4.8 per cent growth in volume and 2 per cent growth in unit prices). The trends for Egypt were similar. Export volumes and values were falling in traditional crops, and unit prices were stagnating. Meanwhile, the annual export value of high-value products was increasing by 6.6 per cent (a 5.5 per cent increase in volume and a 1.1 per cent increase in unit prices).

The predominately large Egyptian commercial farmers have benefited from the development in exports of high-value products. There is a growing consensus that, if smallholders become involved in the production of high-value crops for export, this would be the most effective way of stimulating the rural non-agricultural economy towards positive growth. This study therefore explores the potential of involving smallholders in the production of non-traditional high-value horticultural crops for export, within the thematic context of contract farming, with the aim of improving the return to smallholders on production for the domestic market.

Findings

The study finds that contract farming could be an effective way of including smallholders in the effort to supply the horticultural export value chain, particularly if farmers are organized into farmer associations. The study reveals that smallholder families could increase their incomes by as much as 63 per cent if they engage in the contract farming of organic horticultural produce, and by 43 per cent if they engage in conventional export crops. It is estimated that if farmers would organize into farmer associations they could improve their incomes by curtailing the number of traders and selling their own production in village, governorate, and metropolitan wholesale markets, improving their household incomes by 7, 15 and 22 per cent, respectively, in these markets.

Economic impact. The marginal propensity to consume in villages by smallholders has been estimated at around 64 per cent. This provides an economic multiplier of 2.8, meaning that, for every additional Egyptian pound (EGP) earned by a smallholder, an additional EGP 2.8 would be generated in the local economy. This would drive economic growth in the agricultural sector. Equally important, it would fuel non-agricultural small and medium-sized enterprises in villages, which would generate needed job opportunities. If smallholders produce the crops most suitable for them, this could generate an annual recurrent return (see table below).

- **Civil Society**

  Farmer associations. Community development associations or farmer associations appear to be good starting points for contract farming. However, it is recommended that development efforts should assist smallholders – after they have gained the initial experience of working as a group – in forming independent, registered farmer associations and cooperatives specialized in horticultural crops. This would ease the logistical problem of GlobalGAP auditing and organic certification and make it easier to integrate farmer associations vertically with exporters and for farmer associations to obtain credit.

  Professional associations. The Government should establish a competitive grant scheme open to professional associations and their members. The scheme would support export-promotion activities that would widen exports. Exports are now narrowly concentrated on only a few countries in the European Union.

  Research and development. The Government should establish a competitive grant to support public-private research partnerships for the development of new varieties and new post-harvest techniques and testing, and to modify and adapt small drip-fertigation systems able to cater for horticultural crops and other crops on plots of 1/2 to 1 feddan.

Economic Impact of Smallholder Involvement in the Production of High-Value Export Crops

Production method	Annual incremental feddan needed for export	Smallholders: Number benefiting	Smallholders: Total household return, EGP million	Direct employment creation equivalent, annual full time jobs	Village-leve economic multiplier EGP millior
Conventional	12 700	21 910	49	1 345	15C
Organic	2 800	5 000	19	90	6C

Conclusion and recommendation

To provide an enabling environment for contract farming and improve the competitiveness, handling and transport of the increasing volume of horticultural produce for export, the study finds that the following recommendations would be helpful:

- **Policy issues**

Contract enforcement. New institutional arrangements should be developed that are able to enforce disputes between farmer associations and exporters. Such institutions would contribute significantly to the rapid development of contract farming and the establishment of farmer associations.

Land tenure. Currently, land-rental charges mirror the value of the crops being produced, thus exploiting those farmers taking additional risks in producing high-value crops. This is particularly harmful to landless smallholders, who frequently do not have income-generating alternatives to their rented land. To protect tenants, particularly landless smallholders, a land rental system should be developed that is based on the productivity of the land rather than on the effort and risks taken by tenants in growing high-value crops.

Air cargo-handling. The framework for horticultural cargo-handling at Cairo International Airport has reached its limit in export volume. Moreover, stakeholders believe that air freight is 30-50 percent more expensive there than at airports in competing countries. Only one company is licensed to handle horticultural produce. This is believed to be the main cause of the relatively high freight costs. Another limiting factor is the restriction on the hiring of cargo planes directly by exporters. To improve handling and address the growing volume of air cargo freight, the Government should allow competition among cargo-handlers and allow exporters to hire cargo planes directly.

Investment incentive system. To reduce the enormous post-harvest losses in horticultural produce, which amount to the equivalent of 11 per cent of Egypt's total plant production, the Government of Egypt should develop a tax-incentive package that encourages businesses to invest in cold-chains for horticultural produce.

Government involvement in professional associations. The Government's support for the Union of Producers and Exporters of Horticultural Crops is viewed as a conflict of interest because the union is perceived as a government apex institution for all private horticulture professional associations, as well as for farmers, processors and exporters. This places a damper on the activities of purely membership-based associations. The Government should either make the union part of the Ministry of Agriculture and Land Reclamation or provide purely hands-off support that cannot be misconstrued as a means of controlling professional associations.

- **Investments**

Seaport upgrading. If the port of Alexandria is to keep up with the increased refrigerated sea container freight, it needs to be upgraded. With the expected increase in production and exports of horticultural produce from Upper Egypt, it will become attractive to use the Red Sea ports; however, if this is to happen, the capacity of the ports in handling refrigerated sea-containers also needs to be upgraded.

Smallholders. Development efforts in providing financial services for smallholders should include the financing of:

- production inputs;
- row and high tunnels;
- drip-irrigation equipment suitable for a smallholder;
- field packaging–charcoal cold-storage sheds, and
- pre-cooling and refrigerated transport for larger farmer associations and cooperatives.

Small and medium-sized enterprises. Development efforts to provide financial services for small and medium-sized enterprises should focus particularly on start-ups and the expansion of small and medium-sized enterprises situated in rural areas for investments such as:

- refrigerated cold-chains;
- processing facilities;
- facilities producing packaging materials;
- professional nurseries; and
- business development services.

Organic farming. Organic farming has huge potential, but the development of the subsector would benefit if exporters and processing facilities were situated closer to smallholders to ensure that technical issues are addressed and that produce without an export market is sold to supermarkets and the tourist industry.

Source: Adapted from **IFAD**. 2008. Egypt: Smallholder contract farming for high-value and organic agricultural exports. Pages 6-9. http://www.ifad.org/pub/pn/egypt.pdf

Assessment Test for Tool # 7: **Contract Farming**

Several farmers in your area are interested in managing risk in their farming operations. They have heard of "contract farming" as a promising ARM tool but they know that there are different types of contracts. Considering each farmer's objectives and desires, you, as a farm management advisor, have to tell them which one of the following four contractual arrangements is the most appropriate one for their particular situation.

A. **Marketing contract,**
B. **Production contract,**
C. **Out-grower scheme**
D. **Collective marketing contract**

Situation 1: Some farmers wish to produce their own crops but secure production inputs. They would deliver the quality and quantity of their produce to the processor and would get compensation for this production.
Situation 2: Some farmers wish to establish a guarantee of selling their produce to a major food marketing outlet in the area at a specific price (or formula for determining the price) for a commodity before the commodity is ready to be marketed.
Situation 3: Some farmers wish to join a farmers group as Cooperative and secure contracts with input suppliers and output users to get the best possible deals and guarantees using the relatively large bargaining power of the group.
Situation 4: Some farmers wish to secure guaranteed market access and coverage of some or all non-labour costs. They would simply use their labour to produce a specialised standard commodity according to the plans and technical advice/inputs from a large processor.

Answers

Situation 1: Production contract,
Situation 2: Marketing contract,
Situation 3: Collective marketing
Situation 4: Out-grower scheme

6. References

1. **ActionAid.** 2015, Mar. *Contract farming and out-grower schemes Appropriate development models to tackle poverty and hunger?* http://www.actionaidusa.org/sites/files/actionaid/contract_farming.pdf

2. **Antonaci L., Demeke M., Vezzani A.**,2015. The Challenges of Implementing Price and Production Risk Management in Sub-Saharan Africa, Scientific Paper 9B, ULYSSES project, EU 7th Framework Programme, Project 312182 KBBE.2012.1.4-05, http://www.fp7-ulysses.eu/ , 35 pp.

3. **ABD Institute.** 2008, Jul. *Overview of Contract Farming in Thailand: Lessons Learned*, by Sriboonchitta, S., and Wiboonpoongse, A. http://www.adb.org/sites/default/files/publication/156751/adbi-dp112.pdf

4. **Barrett, C. B., Bachke, M. E., Bellemare, M. F., Michelson, H. C., Narayanan, S., & Walker, T. F.** 2012. *Smallholder participation in contract farming: Comparative evidence from five countries.* World Development,40(4), 715-730. Smallholder Participation in Contract Farming: Comparative Evidence from Five Countries.pdf

5. **Bijman, Jos.** 2008, May. *Contract Farming in Developing Countries: an overview.* http://www.wageningenur.nl/upload_mm/5/c/b/79333121-6f4b-4f86-9e8e-0a1782e784d6_ReviewContractFarming.pdf

6. **DFID**, 2014. *Agriculture and the Private Sector.* https://www.gov.uk/government/uploads/system/uploads/attachment_data/file/318353/AG_and_private_sector__final_.pdf

7. **FAO.** 2013. *Contract farming for inclusive market access,* by Da Silva, C.A., Rankin, M. FAO Rural Infrastructure and Agro Industries Division. Workshop. http://www.fao.org/3/a-i3526e.pdf

8. **FAO.** 2012. *Guiding principles for responsible contract farming operations.* Workshop. http://www.fao.org/docrep/016/i2858e/i2858e.pdf

9. **FAO.** 2011. *Contract Farming: Partnerships for Growth*, by Eaton, C. and Shepherd, A. FAO Agricultural Services, *Bulletin 145*, Rome. Workshop. http://www.fao.org/docrep/014/y0937e/y0937e00.pdf

10. **FAO.** 2005. *The growing role of contract farming in agri-food Systems development: drivers, theory and practice,* by Da Silva, C. A. B. Rome. Workshop. http://www.fao.org/fileadmin/user_upload/ags/publications/AGSF_WD_9.pdf

11. **FAO.** 2002. *Overview of smallholder contract farming in developing countries*, by Simmons, P. Rome, Italy. http://www.ruralfinanceandinvestment.org/sites/default/files/1126800786652_contract_farming_overview.pdf

12. **IFAD and FAO.** 2012. *Good practices in building innovative rural institutions to increase food security,* by Herbel, D., Crowley, E., Ourabah Haddad N., Lee, M. Rome, Italy. http://www.fao.org/docrep/015/i2258e/i2258e00.pdf

13. **IFAD.** 2008. *Egypt: Smallholder contract farming for high-value and organic agricultural exports* http://www.ifad.org/pub/pn/egypt.pdf

14. **Institute for the Study of International Development.** 2012. *Contract Farming as a Tool for Poverty Reduction in Africa* by Adjognon, S., Naseem A. Policy Brief No. 4. https://www.mcgill.ca/isid/files/isid/pb_2012_04_adjognon.pdf

15. **Minten, B., Randrianarison, L., & Swinnen, J. F.** 2009. Global retail chains and poor farmers: Evidence from Madagascar. World Development, 37(**11**), 1728-1741.Global Retail Chains and Poor Farmers Evidence from Madagascar.pdf

16. **Felgenhauer K., and D. Wolter.** 2009. *Outgrower Schemes – Why Big Multinationals Link up with African Smallholders.* **OECD** www.oecd.org/dev/41302136.pdf

17. **Prowse, M.** 2012. *Contract farming in developing countries: a review*. Institute of Development Policy and Management, University of Antwerp. Workshop. http://www.afd.fr/webdav/site/afd/shared/PUBLICATIONS/RECHERCHE/Scientifiques/A-savoir/12-VA-A-Savoir.pdf

18. **UNIDROIT, FAO and IFAD.** 2015. *UNIDROIT/FAO/IFAD Legal Guide on Contract Farming.* Rome. http://www.unidroit.org/english/documents/2014/study80a/wg03/s-80a-wg03-17-e.pdf

19. **United States Department of Agriculture** (USDA), Risk Management Agency http://www.rma.usda.gov/

20. **Vavra, P.** 2009. "Role, Usage and Motivation for Contracting in Agriculture", *OECD Food, Agriculture and Fisheries Papers*, No. 16, OECD Publishing.

21. **Will, M.** 2013. *Contract farming handbook: A practical guide for linking small-scale producers and buyers through business model innovation.* GIZ. https://www.giz.de/expertise/downloads/giz2013-en-handbook-contract-farming-manual-low-resolution.pdf

Tool # 8: Commodity Exchanges and Futures Markets

1. What is it? – Concepts and Characteristics

The objective of this tool is to provide a basic understanding of the modern marketing facilities in the form of commodity exchanges and futures markets. Its aim is to explain how this tool can be useful for farmers to manage risk in agriculture. The knowledge of these sophisticated techniques is intended to empower farmers and/or their advisors but also the policy makers to develop the necessary infrastructure where it has great potential.

A commodity exchange is a goods and/or financial market where different groups of participants trade commodities and commodity-linked contracts, with the underlying objective of transferring exposure to commodity price risks (UNCTAD). In other words, commodity exchanges provide a platform for multiple buyers and sellers to buy and sell commodity-linked contracts on the basis of rules and procedures laid down by the exchange.
A large numbers of sellers and buyers place their legally and financially binding offers/orders of a specific quantity and quality of a commodity at the exchange. This eliminates the need to have face to face contact or the physical display/handling of the product unlike in regular open/unorganized markets. This process leads to more accurate pricing and efficient marketing.

More advanced commodity exchanges also handle futures contracts, providing farmers a mechanism to shift risks to entities that are in a better position and are more willing to bear them. Market-based instruments for managing agricultural price risk are a practical and non-intervening alternative for managing commodity price volatility.

Commodity exchanges may cover some or all of the below mentioned trading facilities:
1. A formal spot market with the physical exchange of commodities,
2. A formal spot market with links to warehouse receipts (certificates),
3. Forward Contracts,
4. Futures Contracts, and
5. Futures Option Contracts.

Most commodity exchanges started with spot market trading where delivery and payment takes place shortly after the contract agreement. Physical trading normally involves a visual inspection and is carried out in physical markets such as a farmers' market or through the deposit receipts from a certified Warehouse Receipt System storage facility. However, as concluded by an UNCTAD (2009) study, virtually all developed country exchanges moved towards futures trade over time as their services in physical trade (spot and forward) became superfluous; most of the exchanges that were not able to make this transition, disappeared.

Definitions of the key terms provided by the USDA's Risk Management Agency are as follows:

- A **forward contract** is an agreement between two parties (such as the farmer/seller and someone who buys the products) that calls for the delivery of and payment for a specified quality and quantity of a commodity (such as a particular crop) at a specified future date. The price may be agreed upon in advance, or determined by a formula at the time of delivery or other point in time.

- A **futures contract** is very similar as it is an agreement to buy or sell a commodity of a *standardized amount*[10] and quality during a specific month in the future, under terms established by the futures exchange, at a price established in the trading pit at the commodity futures exchange.

- **A futures option contract** is a contract that *gives the holder the right, although not the obligation,* to buy or sell a futures contract at a specific price within a specified period of time, regardless of the market price of the futures contract when the option is exercised. Options provide protection against adverse price movements.

- **Hedging** is a strategy that uses futures or options contracts to reduce the risk of adverse price changes prior to an anticipated cash sale or purchase of a commodity.

Futures contracts are important to farmers as they secure a price in advance of a harvest, reducing their exposure to market price drops. The advantage of futures contracts, given that they are standardized, is that they can be traded like a stock. And also, that there need not be an actual sale or purchase of goods at the stipulated time. An example of how futures contract works is provided in the box below.

Example of how futures contract works:

At the planting time in April, a corn producer hopes to secure the selling price of the expected yield of 15 000 bushels in the harvest period. At that time the six-month corn futures contract (October contract) is trading at US$2.50 per bushel. He decides to hedge the entire expected yield by selling three April corn futures contracts on CBOT at a price of US$2.50 per bushel (the standard contract size of a CBOT).

At harvest time, the October futures falls to US$2.00/bushel and local cash prices fall to US$1.88/bushel. The producer offsets his futures position by purchasing back three April corn futures contracts, making a profit of US$0.50 per bushel. He sells the corn in the physical market to the local elevator at the cash price. The corn producer's effective selling price, ignoring transaction costs, works out to US$2.38 per bushel (US$1.88 plus US$0.50).

In case futures prices rise to US$3.50 and local cash prices to US$3.38, the producer makes a loss of US$1 (US$2.50 less US$3.50) in the futures market. He sells the commodity for US$3.38 in the cash market, again ignoring transaction costs, his net sales price works out to US$2.38 (US$3.38 less US$1). In either of the conditions his revenue does not change and is constant at US$35 700.

Thus, by hedging with futures, the producer could avoid revenue fluctuation resulting from adverse price movements.

Source: FAO. 2006, Pages 11-12
http://www.fao.org/docrep/016/ap308e/ap308e.pdf

[10] For example, every soybean or corn contract traded on the Chicago Board of Trade (CBOT) is for 5 000 bushels.

In Africa, since 1990, a number of commodities exchanges were established. However, there are only three that are currently in operation, namely in South Africa (SAFEX established 1996 and deals with futures and stock contracts), Malawi (ACE/2004, deals with forward contracts, warehouse receipts) and Ethiopia (ECX/2008, deals with warehouse receipts, spot contracts; the remaining five; from Kenya, Nigeria, Uganda, Zambia and Zimbabwe are not functioning (Jayne et al. 2014).

There are various reasons for their failure, the main ones being an insufficient volume of trade and high operating and transactions costs. There are certain pre-conditions for the success of agricultural commodity exchanges. A World Bank (2011) study based in Latin American countries outlines them as follows:

- Sufficient market size, market liquidity and the minimum volume of contracts
- A clear and transparent institutional and regulatory framework
- Existence of financial intermediaries to share the credit risk
- Committed agribusinesses
- Standardized but differentiated contracts
- Transition from a cash market to a financial market attracting local and international investors.

As in the case of other market and finance based tools, there is a need and potential for public private partnership in designing and implementing programs under this tool as well.

2. Suitability of the tool

Typically this tool is suitable for individual large scale commercial farmers. However, smallholders when organised, they can also participate through their association. Of course, this tool has a limited use for most farmers especially where a commodity exchange does not exist. However, in countries where such marketing services are accessible, it has a potential to modernise farm marketing, improve marketing efficiency and help farmers to manage market price risk.

This tool can be used in combination with the warehouse receipts system tool and other risk management tools such as enterprise diversification, agricultural insurance, microfinance, etc. This is one of the more sophisticated and complex tools and would require sufficient understanding and guidance to make use of it in practice.

Although, in principle, an agricultural commodities exchange (ACE) can handle any commodity, so far, commodities generally traded have been wheat, barley, sugar, maize, cotton, cocoa, coffee, milk products and pork bellies.

3. Advantages

Commodity exchanges can be highly efficient platforms for buyers and sellers to meet, primarily to manage their price risks better and also to improve the marketing of their products (ADB, 2013).

As summarised by Sitko and Jayne (2012), the liberalization of grain and other agricultural commodities marketing in Africa provides an opportunity to organize private institutions such as commodity exchanges which have the potential to bring benefits of this innovation to a large number of stakeholders including:

- Commercial farmers,
- Large grain traders,
- Industrial processors,

- Banks, and
- Indirectly, smallholder farmers

Commodity exchanges offer a platform for competitively matching a large numbers of buyers and sellers. This can lead to the following benefits to the stakeholders from producers to consumers:

- Stimulated market transparency and price discovery,
- Reduced potential for collusion among market actors,
- Reduced price volatility and bubbles,
- More accurate price information provided to all
- Reduced transaction costs by expanding the range of potential trading partners,
- Likely increase in price received by farmers
- Improved farmers' direct access to markets, reducing their dependence on intermediaries,
- Industry approved inspection and quality certification services provided, and
- Contract enforcement and arbitration services provided to protect against default i.e. Insurance against potential the opportunistic behaviour of trading partners.

In addition, Commodity Exchanges that offer futures contract services helps producers to hedge against the price drops/volatility in the future and reduce risk by increasing the transparency of price discovery and the certainty of prices.

Finally, it is suggested that, given the rainfall dependence of agriculture resulting in production fluctuations, a regional commodity exchange would be more beneficial than strictly national ones (ADB 2013).

4. Disadvantages

Recent reviews (e.g. Jayne et al. 2014, Bloomberg 2015) have concluded that the benefits mentioned above are far from being achieved in any African country except for South Africa. Sitko and Jayne (2012) conclude that the majority of African commodity exchanges remain underdeveloped.

According to their analysis, the six main factors that limit the growth and success of commodity exchanges, especially of the ZAMACE exchange in Zambia[11], are:

1. A lack of success in attracting financial institutions' commitment to commodity exchanges,
2. The anonymous nature of trading on a commodity exchange exacerbates the risks associated with contract non-compliance and opportunistic behaviour,
3. The potential for a conflict of interest among brokers,
4. The potential for market manipulation in a thinly traded market,
5. high fixed costs that are imposed on actors trading in a thin market, and
6. the unpredictability of government intervention in cereal markets.

More specifically, a study on the Ethiopian Commodities Exchange (ECX) by Zangrandi (2012) reported that the farmers or primary traders had poor access to the ECX's storage facilities, and that they faced difficulties in assessing the value of their options. From the demand side, the buyers may not feel they are guaranteed a sufficient quantity and quality at prices lower than what they pay unconventional and import markets.

[11] In Zambia the commodity exchange suspended trade on the exchange in 2012. Since the suspension of trade, ZAMACE shareholders have sought to restructure (Jayne et al 2014).

5. Examples/Case studies

CASE STUDY 8: Malawi Agricultural Commodity Exchange (Jayne/MSU)

This case study describes the Malawi Agricultural Commodity Exchange, a modern marketing-based ARM tools established in Africa. The intended primary beneficiary of this tool is the market oriented farmers through their associations, although traders, financiers, insurance providers, and government policy makers are all stakeholders. Although the trade in spot and forward contracts has grown in recent years, significant challenges remain for its viability. The case study very critically reviews the present conditions; the performance of the Exchange so far, outlines the areas of weakness and provides useful suggestions for successful revival.

Malawi

Despite a relatively small agricultural commodity market, Malawi is home to three commodity exchanges vying for market share. These are the Agricultural Commodity Exchange for Africa (ACE), the Auction Holdings Commodity Exchange (AHCX), and the Malawi Agricultural Commodity Exchange (MACE). This section briefly reviews each of these exchanges and assesses their viability as self-sustaining trading platforms.

Agricultural Commodity Exchange for Africa (ACE)

ACE is Malawi's most developed commodity exchange. It began in 2005/06 and offers both spot and forward contracts. It does not trade in futures contract. Thus, all contracts traded on ACE require physical delivery of commodities either at the time of trade or at a specified future date.

ACE has a diversified shareholder structure that includes gran trading firms, food processors, and farmer's associations. This diversified ownership structure is beneficial in terms of building sector-wide confidence in the exchange. In other countries, commodity exchanges have struggled to gain the trust of potential participants due to ownership arrangements that are concentrated in the hands of one segment of the market, such as the trading sector (Sitko and Jayne 2012).

Unlike some African commodity exchanges, such as ECX in Ethiopia, ACE does not own warehousing infrastructure. Instead, it certifies privately owned warehouses, which are bonded and insured, and have the capacity to effectively grade commodities. In total ACE has 155,000 MT of certified storage capacity across the country. This arrangement enables ACE to keep its monthly operating costs substantially lower than exchanges that manage their own storage. This lower overhead enables ACE to achieve profitability through fewer transactions than is the case for exchanges that manage their own storage.

In addition to grading and storing commodities, three of ACE's certified storage facilities also issue warehouse receipts. Receipts issued from these facilities can be collateralized at three different banks in Malawi. These banks are First Merchant Bank, National Bank of Malawi, and Standard Bank. Warehouse receipts have enabled the development of forward contracts on agricultural commodities, where producers or processors are able to borrow against the underlying commodity at an interest rate of around 20 percent, nearly half the commercial lending rate in Malawi. In total, 8.5 billion Malawian kwacha (~$20.4 million) has been traded in warehouse receipts, mostly for soya beans and sunflower. The concentration of warehouse receipt trade in these commodities likely reflects the relatively lower level of price uncertainty of these commodities, relative to more widely grown crops such as maize, which are frequently subject to unpredictable policy induced price movements.

ACE has benefitted significantly from donor investments aimed at supporting the development of the exchange and from trade in Bid Volume Only (BVO) contracts initiated by the World Food Programme (WFP) under its Purchase for Progress initiative. Donor-funded support for ACE includes $635,000 from the EU to support the development of warehouse receipts and $540,000 from AGRA to support farmer and trader sensitization. In total ACE has received approximately $2 million in donor support since 2005.

BVO trade from WFP and some commercial processors has undoubtedly been the major factor driving trade across the ACE exchange. Since 2012 there have been 49 BVOs for maize and 47 BVOs for pulses, totaling 68,832 MT and 40,950 MT in respective volumes. In the absence of these BVOs it is unlikely that ACE would trade in sufficient volumes to remain viable.

That being said, trade in spot and forward contracts has grown over the years of ACE's operations. Trade in these contracts has increased as a result of several important factors. First, ACE does not require a minimum quantity on its contracts, which enables even small volume producers to offer commodity for sale on the exchange. Second, ACE, with donor support, has invested heavily in sensitization campaigns aimed not only at producers, but also traders and processors. By contrast other exchanges, such as ZAMACE in Zambia, underspent on sensitization, which limited the extent to which the value of an exchange was appreciated in the market, even among more sophisticated market actors such as commercial farmers and large-scale processors (Sitko and Jayne 2012). Third, ACE has developed a network of seven market centers across the country where trades can be conducted. This enables producers and traders from across the country to participate. Finally, trade across ACE has benefitted from an improvement in the agricultural policy environment in Malawi that has contributed to increased predictability and a decreased role of the state in the market. In particular, due to financial constraints, Malawi's grain market board, ADMARC, has substantially curtailed its role in output market over the last few years. As Sitko and Jayne (2012) have shown, improved predictability of state action in food markets is a fundamental precondition to the development of commodity exchanges in Africa.

Since January 2013, ACE has conducted 854 spot and forward contract trades amounting to 83,206 MT of commodities. These trades are disaggregated by commodity type in the Table 1 below.

Table 1: Spot and forward contracts traded on ACE January 2013 to March 2014

	Contracts	
	Number of transactions	**Volume MT**
Grains (Maize)	462	53,414
Beans (soya, kidney)	219	6,401
Pulses and sunflower	36	6,257
Groundnuts	11	88

Source: ACE website: http://www.aceafrica.org/market-info.aspx

While this is a small fraction of the total volume of agricultural trade in Malawi, it does represent a moderate shift toward greater levels of formalization in the agricultural markets in the country.

In summary, ACE's success to date owes to four factors.

First, it can rely on certified privately owned warehouses that are bonded, insured, and provide grading services. By doing so, ACE has been able to keep its operating costs substantially lower than exchanges that manage their own storage. Second, ACE benefits from experienced and dedicated management who have taken a number of innovative steps to increase participation on the exchange in its early days, such as not requiring a minimum quantity on its contracts and setting up numerous market centres in different parts of the country. Third, ACE has benefitted from the more limited role of the Malawian government in the maize market in recent years. Fourth, and very importantly, ACE has benefitted from the commitment of the WFP to use the exchange. Without the support of the WFP, which accounts for most of the trade across the exchange, it is not clear that ACE would be able to cover its costs or expand beyond its currently limited role in agricultural markets.

AHCX (Auction Holdings Commodity Exchange)

AHCX was opened in 2013 by Auction Holdings Ltd. (AHL), which has traditionally provided the platform in Malawi for tobacco trading. The motivation for AHL to develop a commodity exchange is not entirely clear, given the fact that two other exchanges were already operating in the country when AHCX was opened. One respondent suggested that it is an effort to diversify the company's economic base away from tobacco as the global market for burley tobacco weakens. AHCX currently offers only spot contracts for the following agricultural commodities: i) non-GMO maize; ii) soybeans (grades 1 to 3); iii) groundnuts (high-protein or high-oil variety, grades 1 & 2); iv) rice (grades 1 and 2, all varieties); and v) sugar beans (red speckled beans, grades 1 and 2).[10] AHCX plans to trade forward options and futures contacts designed to manage trader and farmer risk.

Like ACE, AHCX provides a platform for trade in specified commodities of a designated quality through warehouse receipts. Like ACE, AHCX provides clearing facilities and arbitration for transactions conducted across the exchange. However, AHCX differs from ACE in some important respects. First, AHCX is fully owned by AHL and therefore does not have shareholder representation from a range of stakeholders and firms in the food system.

While it does have trading members from various segments of the agricultural sector, these are not owner members with a vested interest in ensuring that the exchange functions effectively. Second, AHCX owns and operates all of its own storage facilities, similarly to ECX in Ethiopia. This has the advantage of allowing the exchange to maintain direct oversight over all aspects of the trading system, but also leads to very high monthly operating costs. These high costs must therefore either be spread across a large number of transactions or through high exchange commissions on those transactions that do occur on the exchange.

Third, AHCX appears to enjoy a good deal of political goodwill in Malawi. Indeed, the President, Vice President, and Minster of Agriculture and Food Security were all in attendance at the official launch of the exchange. This high level of political visibility may be the result of its relationship to AHL, which commands significant political power in Malawi given tobacco's role as a foreign exchange generator in the country. It is unclear how the relationship between policy-makers and AHCX will play out in practice, but one can envision an Ethiopian-type situation where certain commodities or actors are required to trade over the exchange if AHCX fails to generate organically sufficient trade levels. Finally, AHCX has yet to achieve the same buy-in from financial institutions as ACE to support the financing of warehouse receipts. Currently no financial institutions are listed as potential financers of receipts on the AHCX website. This lack of support may reflect a wait and see attitude by the financial sector as it assesses the capacity of AHCX to manage the stocks underlying the warehouse receipts.

AHCX does not make available data on trade volumes. However, based on a reading of statements issued by the exchange and according to private interviews, AHCX has traded only 150 tons of maize since its inception. According to the Head of Operations for AHCX Davis Manyenje, AHCX has struggled to achieve sufficient supplies of commodities offered for trade on the exchange. This likely reflects a lack of investment into producer sensitization as well as limited interest from private traders to trade across the exchange.

The exchange has been open for 9 months and over that time it has sold only 150 tons of maize. However its impact on the business ecology, which supports farmers and small- scale traders in Malawi, has already been significant. [11]

Malawi Agricultural Commodity Exchange (MACE)

Though operational, MACE does not have specified contracts that it trades or a dispute resolution mechanism for transactions conducted across its platform. According to industry experts in Malawi, MACE is a commodity exchange in name only. In practice it operates more like a market information system.

Notes:

10. AHCX Web Site, see http://www.ahcxmalawi.com/

11. In a recent press release the executive director of the exchange, Mr. Manyenje, stated that: "We are now seeing several financial institutions coming in to support farmers and traders with warehouse receipt financing on commodities other than tobacco, effectively spreading the benefits that tobacco farmers have long been benefiting to farmers of other crops. The same applies for other service providers that are now coming to offer production inputs, confident that the farmers now have a sustainable and assured market structure, where recoveries of lending is almost assured".

Source: Adapted from Jayne et al. (Oct 2014). Agricultural Commodity Exchanges and the Development of Grain Markets and Trade in Africa: A Review of Recent Experience. Pages 17-21. http://fsg.afre.msu.edu/gisaia/Ag_Commodity_Exchange_Report_FoodTrade.pdf

6. Assessment Test for Tool # 8: Commodity Exchanges and Futures Markets

The basic knowledge of the key terms describing different types of trading contracts is essential to understand how a Commodities Exchange works. In an Exchange, commodities are traded (bought and sold) by a large number of buyers and sellers without having to come face to face and thus increasing the market efficiency.

Please fill in the blanks in the following 1 to 4 definitions with the names of the corresponding trading contracts (a to d).

1. A _________________ contract involves commodities trading at an organized market place where goods are bought and sold for cash and delivered immediately. The contracts are binding and effective immediately.

2. A _________________ contract is an agreement between two parties (such as the farmer/seller and someone who buys the products) that calls for the delivery of, and payment for, a specified quality and quantity of a commodity (such as a particular crop) at a specified future date. The price may be agreed upon in advance, or determined by a formula at the time of delivery or other point in time.

3. A _________________ contract is an agreement to buy or sell a commodity of a *standardized amount* and quality during a specific month in the future, under terms established by the futures exchange, at a price established in the trading pit at the commodity futures exchange.

4. A ________________ contract is a contract that *gives the holder the right, although not the obligation,* to buy or sell a futures contract at a specific price within a specified period of time, regardless of the market price of the futures contract when the option is exercised.

Names of trading contracts:

a. Futures contract,
b. Spot market contract,
c. Futures options contract
d. Forward contract,

Answers

1. = b - Spot market contract
2. = d - Forward contracts
3. = a - Futures contract
4. = c - Futures options contract

7. References:

1. **African Development Bank.** 2013. *Guidebook On African Commodity and Derivatives Exchanges* By Cedric Achille Mbeng Mezui, Lamon Rutten, Sofiane Sekioua, Jian Zhang, Max Magor N'diaye, Nontle Kabanyane, Yannis Arvanitis, Uche Duru And Bleming Nekati.

2. **Antonaci L., Demeke M., Vezzani A.**,2015. The Challenges of Implementing Price and Production Risk Management in Sub-Saharan Africa, Scientific Paper 9B, ULYSSES project, EU 7th Framework Programme, Project 312182 KBBE.2012.1.4-05

3. **Bloomberg Businessweek**, April 2, 2015. Trading Floors Can't Feed Africa - Exchanges aren't helping farmers as foreign backers hoped By Alan Bjerga and William Davison http://www.bloomberg.com/news/articles/2015-04-02/africa-s-commodity-exchanges-fail-to-bring-hoped-for-benefits

4. **FAO**. 2006. An introduction to market-based instruments for agricultural price risk management, by Kang, Myong Goo & Mahajan, Nayana. http://www.fao.org/docrep/016/ap308e/ap308e.pdf

5. **Hernandez, Manuel A.; Rashid, Shahidur; Lemma, Solomon and Kuma, Tadesse.** September 2015. Institutions and Market Integration - The Case of Coffee in the Ethiopian Commodity Exchange. IFPRI Discussion Paper 01464.

6. **Jayne, T. S., Sturgess, Chris, Kopicki, Ron and Sitko, Nicholas**. Oct. 2014. Agricultural Commodity Exchanges and the Development of Grain Markets and Trade in Africa: A Review of Recent Experience. Working Paper No. 88. Indaba Agricultural Policy Research Institute (IAPRI). Lusaka, Zambia. http://www.aec.msu.edu/fs2/zambia/index.htm

7. **Rashid, S.** 2004. Agricultural Commodity Exchanges in Developing Countries: A Critical Review of Literature. Unpublished, International Finance Corporation World Bank Group, Washington, DC.

8. **Robbins, Peter.** 2011. Commodity exchanges and smallholders in Africa. Brief Series. International Institute for Environment and Development/Sustainable Food Lab.

9. **Sitko, Nicholas J. and Jayne, T. S**. 2012. Why are African commodity exchanges languishing? A case study of the Zambian Agricultural Commodity Exchange. Food Policy, 37(3), pp. 275-282.

10. **United Nations Conference on Trade and Development** (UNCTAD), 2009. Development Impacts of Commodity Exchanges in Emerging Markets, Adam Gross and Leonela Santana – Boado.

11. **United States Department of Agriculture** (USDA), Risk Management Agency. http://www.rma.usda.gov/http://www.rma.usda.gov/pubs/2011/rma_glossary.pdf

12. **Whitehead, Eleanor,** 2013. *Africa's Agricultural Commodity Exchanges Take Root,* Forbes and African Edition of the Financial Times, August 15, 2013. Last accessed May 26, 2014:

13. **World Bank.** 2011. Agricultural Commodity Exchanges in Latin America and the Caribbean by Diego Arias, Alfredo Ferreira Lamas, and Musa Kpaka. LCSSD Food Papers Series.

14. **Zangrandi, Michele Savini.** 2012, May. Rwanda Commodity Exchange (RCX) http://www.theigc.org/wp-content/uploads/2015/03/Zangrandi-2012-Policy-Brief.pdf

Tool # 9: Warehouse Receipts System

1. What is it? – Concepts and Characteristics

A warehouse receipt system (WRS) or an electronic warehouse receipt system (eWRS) is a relatively modern innovation based on a variation of traditional storage strategies. It seems to have significant potential and has already been adopted in several developing countries. The main objective of this tool, therefore, is to provide a basic understanding of the concept and workings of WRS, to explain how and where it is applicable and to explain its pros and cons.

A WRS is a formal agreement between a licensed storage facility operator and a named depositor specifying the quality and quantity of a specified commodity held in a secured storage environment. The document, called a warehouse receipt (a certificate of deposit and ownership of the stored commodity) can then be used as a collateral instrument to obtain financing by the depositor from a lending institution or input suppliers.

Besides obtaining credit, the warehouse system aids farmers in making marketing decisions such as how long to store the commodity and how and when to market it. In some countries, WRSs are also tied to commodity exchanges as a physical deposit/delivery system. While it can be considered finance related tool, in this course, WRS is included as a market related tool.

Simple steps in a typical WRS are:

1. A producer deposits agricultural goods in a warehouse
2. The warehouse issues a receipt to the producer
3. This receipt can then be used as collateral to obtain a loan or to market the commodity as desired.

As described in an AFD/CTA/IFAD (2014) study, in SSA there are mainly four types of warehousing services currently in practice:

Type A: Community inventory credit – Under this system, stocks are held in small community warehouses controlled jointly by the producer organisation (or a group of farmers) and the micro-finance institution (MFI) involved. The operations of MFIs are generally re-financed by commercial banks. This system is used by smallholder farmers primarily in Madagascar under the name of *Greniers Communautaires Villageois* (GCVs) and in a few francophone countries of West Africa under the name of *warrantage communautaire*.

In the majority of cases, the individual depositor's stock identity is preserved and once loans are repaid, the owner is responsible for disposing of the product. According to the study, the reported repayment rates are near 100%. The GCVs in Madagascar account for a large part of the lending portfolio of the local MFIs as they handle annual deposits of 100,000 to 120,000 tonnes including paddy rice, milled rice, cloves and coffee.

Type B: Private warehouses. In these cases, the stocks are held in a private warehouse and are under the control and responsibility of a collateral manager (CM). Services of this type of warehouse are available to private users and not for any public depositor. This can include a field warehouse, where the goods are held in the borrower's store, which is temporarily leased to the CM. These are currently being applied mainly to internationally traded cash crops such as cocoa and cotton handled by large traders. This system is well adopted in countries such as Côte d'Ivoire, Cameroon, Burkina Faso and Uganda. Typically, given the high costs of CM, farmers or small producer organizations do not engage in this type of warehouse service and financing.

Type C: Public warehouses. These types of warehouses are typically owned and operated by a private sector, but their services are available to the general public. This type of WRS is currently practiced in countries including Côte d'Ivoire, Ghana, Mozambique, Senegal and Uganda. Several other countries in Africa (e.g. Ethiopia, Madagascar, Tanzania, and Uganda) are currently working on the legislation to establish public warehouses. According to the AFD/CTA/IFAD study, public warehouses also serve as delivery locations for commodity exchanges and they are vital to the establishment of these exchanges.

Uganda is the only country to date to have passed enabling legislation for public warehousing. The country has also installed an electronic warehouse receipt system (eWRS). The study reports that at least two of the five warehouses that were initially licensed remain operational. Obtaining the required quantity was the main cause for failure.

Type D: Lending against the security of current or future production – Under this system the financiers take security over paper (e.g. bonds) as opposed to the possessory collateral of goods held in warehouses. This system is currently not applied in Africa, however, it is popular in Brazil.

This tool is ideal to forge effective public private partnerships in designing and implementing underlying activities.

2. Suitability of the tool

As a tool, WRS is suitable for farmers who produce a marketable surplus of a relatively standardised quality (or produce that is gradable in different grades) of a commodity, provided that a warehouse providing this service is available in the area. By and large, WRS is not widely used by small holder farmers in Africa and other developing regions. They can be made more suitable for smallholder farmers by making them community-based warehouse systems (such as the *Warrantage System* in Niger, developed with the help of development partners).

The WRS is suitable as an ARM tool when -

1. price risk arises from price fluctuation of the commodities produced,
2. there is a financial risk of not being able to obtain credit on time and having to pay off creditors when the market prices are too low, and
3. post-harvest loss risk needs to be reduced by being able to acquire secure and safe storage facilities.

The WRS is especially suitable for grains and legumes where high seasonal price variability offers an opportunity for gains from selling at high prices, usually much later than the harvest or the immediate post-harvest period.

The system is also more suitable for cash crops as it helps to improve transparency in the marketing of the product (for example, of raw cashew nuts). As indicated by the AgriFin (2013) study, almost the entire crop of cashew nuts produced in Tanzania where the WRS is most developed in Africa, is currently marketed through the WRS.

3. Advantages/Opportunities

- A successful WRS can provide benefits for all concerned parties including producers, traders, creditors and warehouses, as well as for the agricultural sector overall (FAO 2015).

- It can potentially provide credit to farmers, producers and traders by solving the major obstacle of the lack of collateral for such loans.
- In addition, storing the produce in a secure environment allows producers to take advantage of price fluctuations and sell when the market conditions become favourable.
- A WRS can also allow smallholder farmers to participate in commodity markets, improve their bargaining power by selling the commodity as a group and fetch better prices (Coulter and Onumah, 2002).
- The system may also help to reduce marketing costs by allowing for independently enforced commodity standards especially when commodity-specific quality standards are used (Onumah et al).
- Given that the lender holds the warehouse receipt and legal rights to seize the stored goods, the loan default or non-payment risk for the lender is also reduced.
- Effective storage with a sufficient capacity will also help stabilize prices by reducing the seasonal variation, thus benefitting consumers.
- Smallholder farmers can benefit from this tool especially when there is strong community involvement bringing in peer pressure as well as an effective legal framework and the rules governing the system to promote accountability and transparency on the part of the warehouse service provider and the lenders.

4. Disadvantages

- Risk of inefficiency, mismanagement, fraud, failure or bankruptcy on part of the warehouse or the MFI.
- It is hard to implement if the network and state of MFIs is weak in the country.
- Lack of expertise for WR service providers to handle the attack of pests or other hazards to crops stored in the warehouse.
- Lack of suitable storage infrastructure in local areas.
- Absence of an enabling legal and regulatory framework and governing rules.
- Costs for storage facilities and related services.

5. Examples/Case studies

CASE STUDY 9: Empowering farmers in Tanzania through WRS (IFAD)

This case study explains how the IFAD-supported WRS market development program is helping farmers in Tanzania to build their resilience to better cope with risks they face in farming. The key stakeholders are the farmers, the government, policy makers, warehouse service providers, traders, bankers, IFAD and the international donors. The report shows that the capacity of small farmers to manage risk through a secured access to credit and storage facilities can be enhanced by this tool.

Empowering farmers in Tanzania through the warehouse receipt system

When farmers have secure access to credit and reliable storage facilities for their grain, it gives them the option to sell when they can get the best price. This means that in a situation of rising food prices small farmers stand to benefit, not to lose. The warehouse receipt system, introduced through the IFAD-supported Agricultural Marketing Systems Development Programme in Tanzania, is now being mainstreamed by the government throughout the

The harvest has been good. It's the end of a busy day and Maimuna Omary Ikanga, a farmer from Qash in the Babati district of Tanzania, has loaded an ox-drawn cart with sacks of grain harvested on her land. She sets off to the warehouse where attendants weigh the grain, measure moisture levels and stitch up the sacks for storage. They give Maimuna a warehouse receipt for her grain.

Maimuna Omary Ikanga weighs her maize in the Qash receipt system warehouse

Credit: IFAD/M. Millinga

That receipt is no ordinary piece of paper. It represents an opportunity for Maimuna to grow as a small-scale entrepreneur and continue to lift herself and her family out of poverty. With the receipt, she can use her stored grain as collateral to get credit at reasonable interest rates, and she can continue to build the small business she started in the early days of the receipt system.

"Before, I was harvesting my crop and selling it at a low price, which was all I could get," says Maimuna, who grows maize, peas, beans and sunflowers. "Now, if I store it, get a loan and wait for prices to go up, I can make a profit."

Linkages are key to success

The warehouse receipt system is the result of a collaboration between two IFAD-funded projects that are being pioneered in the Babati district. The success of the system relies on a series of linkages that are being addressed by IFAD's projects in Tanzania. They involve access to storage facilities, credit, markets and market information.

The Agricultural Marketing Systems Development Programme (AMSDP) was set up in 2002 to improve the structure and performance of Tanzania's crop marketing systems. The seven-year programme, which ends in 2009, is working in four main areas:

- developing agricultural marketing policy
- empowering small producers by building their entrepreneurial and organizational capacity and improving their links to markets
- providing marketing-related financial services so that small farmers can secure loans to cover the period between harvest and sale
- developing rural marketing infrastructure, including storage facilities, marketplaces and roads

Amina Isare and Sellemn Ally treat maize with pesticides at the Qash receipt system warehouse

Credit: IFAD/M. Millinga

The infrastructure component of the AMSDP has built safe, managed storage facilities for farmers that fulfil all the requirements for maintaining the quality of the product.

The Rural Financial Services Programme supports the creation of savings and credit cooperatives (SACCOs) formed by local communities, which allow poor rural people to get much-needed credit at reasonable rates. The warehouses have been built or rehabilitated in areas where SACCOs have considerable experience. Once the harvesting season begins, SACCOs managers submit a loan application to the bank. The warehouse manager issues a receipt to the farmer when the produce has been deposited in the warehouse. The farmer can use this receipt to obtain a loan from the SACCOs of up to 70 per cent of the value of the deposited stock. Farmers can then wait for better prices before they sell.

Maimuna Omary Ikanga and helpers take her produce to the receipt system warehouse in Qash, Babati district

Credit: IFAD/M. Millinga

"The warehouse receipt system is an arrangement that solves two problems: the lack of storage facilities in the district and the difficulty of obtaining credit," says Vincon Nyimbo, marketing specialist of the AMSDP. "At certain times of the year farmers need cash for various reasons. These warehouses help manage the food security issue and the marketing issue."

Historically banks have been reluctant to finance agricultural-related activities. The uncertainty of external factors such as drought or floods makes investment too risky. The warehouse receipt system provides a way of getting around that.

Crop prices usually decrease drastically during the harvest season, but after three to six months the prices may double or triple. Farmers without storage are forced to sell their produce when market prices are low. Traders often exploit this situation. They have storage facilities and can sell with a good profit margin once market prices improve.

The links to markets and to up-to-date market information forged by the AMSDP and the First Mile Project, which is supported by the Government of Switzerland, are a third factor in the success of the warehouse receipt system. Farmers in Babati, for example, are well informed of movements in the markets and they are able to access markets rapidly, so they can wait to sell at the right time for the best price.

Profits for investment

Labels identify stored rice in the Magugu warehouse

Credit: IFAD/M. Millinga

The warehouse receipt system has had an immediate and positive effect on farmers' incomes. Some have been able to use the credit to venture into new enterprises.

In 2006, when the harvest was good, Maimuna and others nearly doubled the income from their produce. "We kept it in the warehouse when the price was 15,000 Tanzanian shillings (US$13) for a 100 kilogram sack and we sold for 26,000 shillings a sack when the price rose."

Maimuna invested her profits in a satellite dish, a television set and a generator. With the nearest electricity source 10 kilometres away, this was a major attraction for the community.
"I bought it to watch the World Cup," she says. She charged fellow villagers to watch the football matches on her television set. "To tell the truth, I was making money! In just a day I was making 18,000 to 20,000 shillings. And it went on for a month..."

Maimuna is now building a second house. If harvests continue to be good, perhaps she will be able to finish it next year. Maimuna has also become a SACCOs chairperson. She is responsible for organizing meetings and represents members of her association at the meetings.

More and more farmers have been using the system. As a result, the SACCOs have been able to obtain bigger loans on more favourable terms, which in turn has had a positive impact on farmers' incomes. Farmers in other areas are demanding that the system be extended. Outside the programme area some farmers' associations have even been implementing warehouse receipt systems with their own funds.

In view of the success of the system in Babati and other districts, the government wants to extend it throughout the country.

"We are in the process of refining the system so that we can give practical recommendations to the government as it replicates the system," says Nyimbo. "It was important, for instance, that there was a legal framework for the operation of the warehouses. The Warehouse Receipt System Act of 2005 allows the private sector to own and manage warehouses. This is an excellent example of how programmes like this can have a direct impact on policy development within the country. The Act was an answer to needs identified while piloting the warehouse receipt system."

Source: IFAD http://www.ruralpovertyportal.org/country/voice/tags/tanzania/warehouse

6. Assessment Test for Tool # 9: Warehouse Receipts System

Match/pair the following "Type of Warehouse System" with its corresponding description/definition.

Type of Warehouse Receipts System:

Type A: Community inventory credit
Type B: Private warehouses
Type C: Public warehouses
Type D: Lending against the security of current or future production

Description:

1. In these cases, the stocks are held in a private warehouse and are under the control and responsibility of a collateral manager (CM). Services of this type of warehouses are available to private users and not for any public depositor. This can include a field warehouse, where the goods are held in the borrower's store. Suitable for internationally traded cash crops such as cocoa and cotton handled by large traders; examples include Côte d'Ivoire, Cameroon, Burkina Faso and Uganda.

2. Under this system, stocks are held in small community warehouses controlled jointly by the producer organisation and the micro-finance institution (MFI) involved. This system is used by smallholder farmers primarily in Madagascar under the name of *Greniers Communautaires Villageois* (GCVs) and in a few francophone countries of West Africa under the name of *warrantage communautaire*.

3. Under this system, funds are provided against the security of a current or future production of the farm. Example - the *Cedulas de Produtos Rural* (agricultural bonds) system that is popular in Brazil. This system is currently not applied in Africa as it requires documented security of current or future production.

4. These types of warehouses are typically owned and operated by a private sector but their services are available to the general public. This type of WRS is currently practiced in countries including Côte d'Ivoire, Ghana, Mozambique, Senegal and Uganda. These warehouses also serve as delivery locations for commodity exchanges and they are vital to the establishment of these exchanges.

Answers

1. = Type B: Private warehouses
2. = Type A: Community inventory credit
3. = Type D: Lending against the security of current or future production
4. = Type C: Public warehouses

7. References:

1. **Agence Française de Développement (AFD), Technical Centre for Agricultural and Rural Cooperation ACP-EU (CTA), International Fund for Agricultural Development (IFAD).** (September 2014). Study on appropriate warehousing and collateral management systems in Sub-Saharan Africa. Volume I – Key Findings, Volume II – Technical Country Reports, and Volume III – Review of Applicable Laws by J Coulter Consulting Ltd., Sullivan & Worcester UK LLP.

2. **Antonaci L., Demeke M., Vezzani A.**,2015. The Challenges of Implementing Price and Production Risk Management in Sub-Saharan Africa, Scientific Paper 9B, ULYSSES project, EU 7th Framework Programme, Project 312182 KBBE.2012.1.4-05, http://www.fp7-ulysses.eu/ , 35 pp.

3. **Coulter, J. & Onumah, G.E.** 2002. The role of warehouse receipt systems in enhanced commodity marketing and rural livelihoods in Africa. Food Policy 27.

4. **FAO.** 2015. Designing warehouse receipt legislation: Regulatory options and recent trends By Philine Wehling and Bill Garthwaite, FAO Investment Centre, Rome, Italy

5. **FAO.** 2009. The Use of Warehouse Receipt Finance in Agriculture in Transition Countries, Working paper by F. Höllinger, L. Rutten, & K. Kiriakov. Rome.

6. **Onumah, Gideon E. (April 2013).** Warehouse receipt financing in agriculture in Africa. **AgriFin** https://www.agrifinfacility.org/warehouse-receipt-financing-agriculture-africa

7. **World Bank.** 2006. Expanding Post-Harvest Finance Through Warehouse Receipts and Related Instruments by Baldwin, M., Bryla, E., & Langenbucher, A., Washington D.C.

LESSON 4
Government-Based Agricultural Risk Management Tools

Learning objectives

At the end of the lesson, learners should be able to understand and argue about different agricultural risk management policies that can be implemented with public support, appreciate the opportunities/advantages/benefits, challenges/disadvantages/costs, the complementarities between policies and tools, and judge the suitability of them in a given area or farm situation.

Although there are significant implications for farmers in managing the risk of their activities, the primary audience of this lesson is on public policy makers. Governments have a role to play in the country's economy, especially in correcting market failures that are primarily due to the existence of information asymmetries and high transaction costs due to the under-provision of public goods. Well-designed Government policies can potentially improve resource allocation, market efficiency and equity through redistribution and positive intervention approaches in terms of market creation, improving market incentives, reduction/mitigation and coping with risk (OECD 2009).

To achieve these goals, governments provide agriculture and food related assistance which may have significant implications for risk management in agriculture. The following three main public support programs are covered in this lesson (note that this is not an exhaustive list) which may have the capacity to alter the nature of risk in agriculture and/or the risk behaviour of decision makers.

Tool 10: Public Foodgrain Reserves
Tool 11: Disaster Assistance Programs
Tool 12: Social Protection and Productive Safety Nets

Historically, the involvement of governments in managing food reserves, either for the purpose of price stabilization or emergency food security, has been very common, in both developed and developing countries. However, corresponding with relative global food shortages and abundance, the overall interest and nature of public involvement in food reserves for price stabilization has been cyclical over time. In general, a more market oriented approach to public support has gained momentum, reducing the need for a direct control of the supply and/or demand of commodities.

Today, the main challenge of foodgrain reserves is how to maintain the necessary level of national food security at a reasonable cost and economic efficiency. The set of policies followed for this purpose have significant implications to the stability of markets and prices for farmers, governments and consumers and thereby for the level of risk and risk-behaviour of farmers.

Disaster assistance, either as a measure of prevention or mitigation, is also employed in many countries, especially by those that are frequently affected by natural disasters. There is a large role for governments to play in dealing with the reduction and management of risk caused by disasters. In some cases, these programs may act as free insurance protection or as an aid to stabilize the livelihoods of those affected, usually poor and vulnerable farm families. This then can have significant implications for those farmers who intend to receive such benefits by way of modifying the risks they face, their own risk behaviour and their management strategies.

The third type of program in the public support category is the provision of social protection (including cash or non-cash transfers aimed at reducing poverty and vulnerability) and agricultural productive safety nets (including farm input subsidies aimed at improving the food security of vulnerable farm households). These social safety net programs have grown in popularity and can be found in almost all countries. The main challenges for policy makers with these public support programs include budgetary implications, targeting such programs and the effectiveness in achieving their goals.

Finally, similar to other lessons this lesson also does not cover all possible tools. For example various government schemes, community sponsored programs, development agencies/INGO subsidies and other assistance programs are not covered explicitly.

Key words for lesson 4.

1. *buffer stocks,*
2. *cash transfers*
3. *commodity stabilization fund,*
4. *disaster assistance,*
5. *disaster risk management (DRM),*
6. *disaster risk reduction (DRR),*
7. *emergency reserves,*
8. *farm input support programs,*
9. *human-induced disasters,*
10. *natural disasters,*
11. *price ceiling,*
12. *price floor*
13. *productive safety net,*
14. *social protection,*
15. *subsidy*

Tool # 10: Public Foodgrain Reserves

1. What is it? – Concepts and Characteristics

Many developing countries since the 1970s, especially after the 1974 food crisis, have followed storage or stock policies to help control unfavourable price swings and to mitigate the adverse impacts on farmers as well as on low income consumers. The focus of this tool, therefore, is on briefly explaining the nature of these policies, how they are supposed to work, what the experiences of various countries have been and what is the relevance of these polices for risk management in agriculture today. The advantages and disadvantages of maintaining such public foodgrain reserves and the key challenges or issues related to this policy are also discussed. Given that the aim of this policy is to reduce the price and market risk of selected agricultural commodities, and in general improve farming resilience, this policy acts as a potential tool for ARM.

There are two main types of public stocks or reserves:

1. Food reserves for price stabilization, and
2. Emergency food security reserves

Food reserves for price stabilization (also called strategic grain reserves or buffer stocks) are aimed at maintaining a reasonable price band with a price floor and a price ceiling. Typically, the main, staple foodgrains are covered for this purpose, although in principle any important commodity from a domestic food security point of view can be procured and stored. Usually, the government grain marketing agency or national food reserve organization handles the function of procurement and distribution. In general, stocks are procured and set aside when the supplies (domestic production and imports) are relatively high and prices are plummeting. Stocks are released (sold or distributed) when the supplies are too low, causing prices to skyrocket.

When a commodity price is stabilized within a certain band (especially if this is also a well-publicised government policy), it reduces the uncertainty for farmers and helps them to manage the price risk. However, from the farmers' point of view it is the minimization of falling prices (down side risk) through the policy of public buffer stocks that is particularly important. In order to maintain a price floor, the government has to declare the minimum procurement price and be ready to buy and store a commodity at that price, especially when the open market prices fall below this floor. The guarantee of a minimum price, especially if it is set at a reasonably remunerative level and is announced well in advance before crop planting, helps farmers to eliminate some of the risk associated with market access and price drops. Thus, in theory, this policy can be an important ARM tool. In practice price policy decisions can become a source of uncertainty and is hard to find truly success examples.

A price stabilization band is determined on the basis of the estimated costs of the policy that are politically and fiscally acceptable to the government. In general, the tighter the band, the more expensive it is to maintain. The floor may be related to the export parity price and ceiling to the import parity prices (AFD 2014). In some countries, the floor price is determined by the minimum support price (MSP) for farmers and the ceiling may be determined by adding a reasonable marketing margin so as to allow the private traders to operate in the market and more or less within the limits of the band. Local traders may be able to buy the commodity from farmers at the MSP or slightly below, by providing additional incentives such as immediate cash payment or, a more convenient point of delivery, etc.

An alternative to actual procurement and maintaining physical stocks is holding a monetary **commodity stabilization fund**. However, to be effective, such a fund has to be very large or the country needs to have ample access to foreign borrowing. Given the practical difficulties of setting aside a sizable financial reserve, this option has not been widely practiced. Furthermore, given the poor performance of stabilization schemes over the years, the emphasis now is more on policies that help manage the risk (such as insurance, organized and transparent commodity markets, futures markets, warehouse receipt systems, etc.) instead of managing agricultural markets through active involvement in manipulating supply and demand (FAO 2006). However, many developing countries find a combination of several policies including price stabilization stocks as a socio-politically preferred course of action.

In addition to the public reserves, foodgrain stocks can also be held by the private sector with subsidies from the government. However, the effectiveness of this alternative in achieving the final goal of controlling prices, (especially near the government policy stated upper limit) has been questionable and in some cases counter-productive due to inherent private business tendencies of hoarding for speculative profits (Galtier 2014, Timmer 2014).

Emergency food security reserves – With or without price stabilization stocks, many countries maintain food reserves with the primary goal of helping the vulnerable population to cope with high market prices by providing foodgrains at low subsidized prices during periods of crisis. Such stocks when released in an emergency situation, may indirectly have some dampening effect on prices depending on the quantity of the food distributed to the consumers, however, a well targeted distribution would be effective in reaching the poor without having a big effect on the market prices.

Provided that vulnerable and disaster-affected farm families are included in the list of beneficiaries of emergency food distributions, this policy, similar to the disaster assistance or social safety net, can also be treated as helping such families to manage production and price risk in agriculture. It should be noted that most smallholder farmers in developing countries are net buyers of food and are equally affected by high prices and food shortages. Thus an added cushion for farm families helps manage family livelihood, farm input purchases and investments to regenerate agriculture following a disaster.

According to Antonaci et al (2015) in Africa, **Kenya** normally maintains a limited amount of Strategic Grain Reserve stock of up to four million bags (expected to be raised to eight million bags soon). **Malawi** also has had a history (since 1999) of strong price stabilization as well as food security reserve programs. In **Ethiopia,** emergency food security reserves have ranged from about 200,000 tonnes in the 1980s and 1990s to about 400,000 tonnes in the early 2000s and are under consideration to be increased to 1.5 million tonnes. The Southern African Development Community (**SADC**) has plans to set-up a regional strategic grain reserve to help bail out countries experiencing food shortages and discourage member countries from imposing restrictions on maize trade within the region. According to the plan, a 500,000 ton facility, including a cash component for countries that do not have any surplus to contribute, would be set up.

2. Suitability of the tool

This tool is suitable in countries where local governments are highly concerned and sensitive to price volatility and food insecurity. Many populous countries in Asia, including China, India, Thailand, and Vietnam among others, follow government procurement policies at pre-set minimum support prices, particularly for rice. Their experience demonstrates that rice prices domestically can be kept reasonably stable with respect to world prices, with the help of government stock interventions

(Dawe and Timmer 2012). However, these national policies may in fact exacerbate the impact on international prices as was experienced from the 2008 food prices crisis.

The policy of emergency food reserves is suitable where a market failure in assuring short-term food security becomes a major concern. A review of stock policies of Southeast Asian countries suggests that there is now a resurgent interest more on emergency reserves and less on the traditional buffer stocks for price stabilization (Briones 2014).

Studies also show that government procurement to build buffer stocks is more effective in controlling the price floor, than it is in controlling the price ceiling (Antonaci et al. 2015). A World Bank study (2012) concludes that "If used, public grain stocks need to be incorporated into a coherent longer-term strategy that combines the use of trade, investments in agricultural productivity, and well-managed, targeted safety net programs."

3. Advantages

From the farmers' point of view, the major benefit of a public procurement and buffer stock policy is the protection from a collapsing market. Depending on the level of the price floor, it may also improve the net income, stimulate production to improve food security and reduce the downside price risk. In many countries, notably for rice in Asia, public procurement programmes also help integrate smallholders into markets. (FAO 2014).

The use of emergency food reserves generally permit a price transmission of higher prices to producers providing the incentive to increased domestic food supply that can subsequently help lower domestic food prices while providing a safety net to the poor, including food deficit farmers. (World Bank 2012).

4. Disadvantages

The major drawback of maintaining food reserves for price stabilization purposes is its costs - procurement, transportation and distribution. The experience of most countries is that these are high and unpredictable. A major study by the World Bank (2012) concluded that the high budgetary costs of price stabilization programs still frequently practiced in Asia and Africa are "crowding out needed public investment in agricultural productivity and rural infrastructure" and discouraged private investment in the grain sector.

Public price stabilization programs with effective price ceilings do not allow farmers to benefit from the high prices even where the price transmission of high international or consumer market prices to the farm gate prices is possible. Similarly, if the prices are capped at a relatively low level, farmers may end up paying for this policy, resulting in negative incentives for agriculture.

The benefits of public procurement at a uniform MSP throughout the country brings uneven benefits to farmers as compared to a no-public procurement scenario. Farmers in surplus production areas benefit more than those in deficit areas. The government has to cover the costs of transporting the grain from production areas to consumption areas. Stocks need to be overturned to maintain the quality of the grain. This adds to the costs of the program even more.

From the review of the literature it can be concluded that with the exception of the case of rice in some Asian countries, there are hardly any examples of effective and efficient price stabilization programs. A study done by Antonaci et al. (2015), lists the following as some of the factors responsible for this limited success:

- The lack of a well-functioning and cost-efficient stock management system;
- the absence of market information systems that rely on solid production forecasts and consistent price analyses; and
- ad-hoc and non-transparent government policy decisions on reference prices and management of stocks.

5. Examples/Case studies

EXAMPLE: Empirical Estimates of Gains from Food Price Stabilization

Most studies have concluded that the efficiency gains from agricultural commodity price stabilization are generally quite small when measured as a proportion of household incomes. They have also concluded that the gains are higher (i) for risk-averse households, (ii) for large net sellers, (iii) when stabilizing prices of staple crops that tend to be more volatile than export crops, (iv) when stabilizing a bundle of multiple products rather than a single commodity, and (v) in countries where foodgrains account for a large share of gross domestic product and consumers' spending.

The first estimate of stabilization gains was undertaken by Newbery and Stiglitz (1981) who found the gains to farmers from complete price stabilization range from 0 to 3 percent of household income, depending on assumptions of risk aversion. Srinivasan and Jha (2001) made similar estimates for the Indian economy as did Islam and Thomas (1996) for five Asian economies, and also found that the static gains to price stabilization were quite small, at about 1.5 to 3.5 percent of farm income. More recently, Myers (2006) applied the same model but distinguished between poor and affluent producers. Depending on the risk aversion of each group, affluent producers were estimated to gain the most from more stable prices (adding 9 percent to their income), while poor producers gained less (3 percent) because their sales were a smaller share of income. Bellemare et al. (2011) analyzed the impact of price stabilization in a multicommodity framework in Ethiopia and found that farmers who have big surpluses can gain from 6 to 32 percent of household income from the stabilization of prices of seven commodities, depending on risk aversion assumptions and whether they produce coffee. At the same time, poorer farmers and consumers gain little and even lose. While the multicommodity integrated framework is superior to one-commodity estimates, it is difficult to imagine full price stabilization of one or two staple crops, much less seven of them. Investment in rural infrastructure to reduce the gap between export and import parity prices and promotion of trade is a necessary and more efficient way to achieve multiple crop price stabilizations, not public stocks.

The vast majority of quantitative estimates of price stabilization have focused on export crops that typically make up a very small proportion of domestic consumer expenditures, such as coffee, cocoa, cotton, jute, rubber, and wool. Studies of food crops are rare, though gains from the price stabilization of these crops are expected to be higher, given their higher price volatility, particularly in Africa, and their high share in household expenditures (World Bank 2006)

It is important to note that the economic gains of stabilization are diminishing as foodgrains become a smaller proportion of value added in developing economies and as a share of household expenditures. An econometric assessment of the 25-year period from 1970 to 1995 in Indonesia showed that rice price stabilization efforts paid very high dividends in fostering economic growth in the first two five-year plans, apart from the additional benefit provided by enhanced political stability (Timmer 1996, 2004). By the mid-1990s, however, as the share of rice in value added and the consumption basket declined, benefits from market interventions diminished, requiring a much more market-oriented rice policy in Indonesia to bring about new efficiency gains.

Source: World Bank. 2012. Using Public Food Grain Stocks to Enhance Food Security. Report Number 71280-GLB. Washington DC. Box 2.1, Page 7. http://www-wds.worldbank.org/external/default/WDSContentServer/WDSP/IB/2012/09/03/000386194_20120903050714/Rendered/PDF/712800ESW0P1130Foodgrain0Stocks0web.pdf

CASE STUDY 10: Regional cooperation in public stocks - Southeast Asia (Bribones)

The principal stakeholders of this tool are government policy makers, farmers and their organizations, people who are expected to be affected by the potential disasters and crises and other consumers, especially in food deficit countries, participating in this scheme. Regional cooperation on public stocks for emergencies is very rare. Thus, even the establishment of the scheme and the agreement on some clear rules for the disbursement of the stocks among 13 countries in the region itself is an achievement. The group has added a humanitarian component to other emergency reserves in recent years. Although the amount of grain stock is relatively small, it is intended to have a positive impact on the recipient country's farmers and their risk bearing ability and capacity to deal with risk in agriculture especially when farm families affected by disasters are also the beneficiaries.

Regional cooperation in public stocks

The roots of regional cooperation

From country-level experience, the discussion now shifts to regional cooperation in public stocks. In Southeast Asia, such regional cooperation was initiated in the late 1970s after the world rice crisis of 1972 – 1974. The ASEAN Food Reserve (AFSR) Agreement was signed in 1979 by the original five member countries, namely Indonesia, Malaysia, Philippines, Singapore, and Thailand. The Agreement establishes the ASEAN Emergency Rice Reserve (AERR), consisting of rice stocks that are earmarked to meet emergency requirements in the region. The earmarks can be part of, or over and above, the national food security reserve. The Agreement does not require distinct physical stocks of rice, as long as the member state makes its earmark available to other members, as a permanent commitment.

Earmarks of the original five signatories are as follows (in tons):

Indonesia:	12,000
Malaysia	6,000
Philippines	12,000
Singapore	3,000
Thailand	15,000
Total	50,000

The original level of AERR earmarks is equivalent to about 8.8% of imports in 1975, or about 0.2% of consumption. [14] The release mechanism is predicated on bilateral negotiation: the country in emergency must direct its request to another member country for release from the earmark of the latter. Other member countries are notified of this request; the demanding and supplying countries then agree on the terms and conditions of release.

Upon accession of additional countries to ASEAN, the AERR also expanded, eventually reaching 87,000 tons, broken down as follows:

ASEAN – 5	50,000
Brunei	3,000
Cambodia	3,000
Laos	3,000
Myanmar	14,000
Vietnam	14,000
Total	**87,000**

The definition of "emergency" in the AFSR Agreement is very stringent: it is *the state or condition in which an ASEAN Member Country, having suffered extreme and unexpected natural or man-induced calamity is unable to cope with such state or condition through its national reserve and is unable to procure the need through normal trade* (Art. IV, Sect. 5).

As seen in the foregoing, the size of the reserve is small; despite continuing growth of imports in the region, the reserve has stayed below 90,000 tons. Moreover the means of accessing the reserve appears to duplicate a normal trade process among member countries, which is government-to-government (G-to-G) trade. Lastly the definition of emergency legally restricts the reserve to be an instrument of last resort, i.e. only when normal trade *and* national reserves are inadequate to meet the shortage. In practice therefore no cross-border releases of rice where ever made by AERR over the next three decades.

The East Asia Emergency Rice Reserve

In 2003, after an evaluation of AERR, an expanded version of regional reserve was initiated, involving the Plus Three countries. This expanded scheme began as a pilot project of the ASEAN Plus Three Ministers of Agriculture and Forestry, called the East Asia Emergency Rice Reserve (EAERR). The project, approved in 2003 and launched in 2005, was implemented until 2010. The EAERR carried over the AERR stocks, and eventually included the earmarks of the Plus Three countries, expanding the regional reserve to 787,000 tons:

ASEAN – 5	50,000
China	300,000
Japan	250,000
Korea Republic	150,000
Total	**787,000**

The EAERR furthermore added a new kind of reserve, which is the stockpiled emergency rice reserve (or stockpiled rice for short). This consists of donated rice stocks administered by EAERR to meet an "acute emergency" in a member country; in practice, this is a humanitarian response in the aftermath of a calamity.

In total, the EAERR released nearly 3,000 tons of rice to households affected by calamities, in Cambodia, Indonesia, Laos, Myanmar, the Philippines, and Vietnam. In 2010, the EAERR also implemented a release of earmarked rice stocks equivalent to 10,000 tons from Vietnam, to address the "lingering effect of calamity" in the Philippines after Typhoon Ketsana.

The ASEAN Plus Three Emergency Rice Reserve

The EAERR pilot was converted to a permanent mechanism called the ASEAN Plus Three Emergency Rice Reserve (APTERR). The APTERR Agreement was signed in 2011, and Agreement entered into force in 2012. The APTERR is to be governed by a Council with day-to-day management by a Secretariat. The APTERR was formally launched in March 2013 with the first meeting of the APTERR Council, composed of thirteen members, or one representatives from each APTERR Party.[15] The Secretariat is based in the host country Thailand.

The APTERR carried over the earmarked and stockpiled emergency reserves of EAERR, as well as the definition of emergency from AERR. Operation of APTERR is supported by financial contributions of APTERR Parties as stated in the APTERR Agreement.

Governance is vested on the Council; day-to-day management is assigned to the Secretariat. Based on the APTERR Agreement, the Council votes by consensus. An APTERR Party is permitted to unilaterally suspend its obligation.

The Council formally adopted a set of internal rules and regulations on establishment and release of APTERR stocks, based on the APTERR Agreement, best practice in emergency reserves, and the experience of the EAERR pilot. Earmarked stocks are now subject to replenishment within one year upon release. The stocks are defined to be milled rice, must be owned and/or controlled by an APTERR Party, and safe for human consumption; there are otherwise no requirements imposed on earmarked rice. [16]

APTERR adopted three programs or "Tiers" by which APTERR stocks could be released. Under Tier 1, earmarked reserves are pre-positioned on standby basis between a supplying and a receiving country in case of emergency in the latter. The standby arrangement already specifies quantity of rice, rice grade, and terms and conditions of release, to avoid protracted negotiations in the even of emergency. The target date of arrival of earmarked stocks is one month or earlier after request by the country in emergency. Contingent delivery is formalized by a forward contract which is valid for three years and renewable. In case earmarked stocks are requested in the absence of a forward contract, release can still be authorized under Tier 2, which is open to bilateral negotiation between requesting and supplying countries. Lastly, release of donated stocks to meet acute (and likely localized) emergency falls under Tier 3.

A particularly contentious issue for release of earmarked stocks is the price. Tier 1 does not require price to be agreed going forward; however it does mandate that the *formula* for determining price be agreed *ex ante*. One formula offered is to simply use the prevailing FOB price of a comparable rice grade under forward contract to be adopted, with predetermined adjustment for quality difference and terms of delivery (e.g. supposing delivery CFR or CIF, as is often the case).

Notes:
14. Hanpongpandh, S., 1982. Modeling the impact of the ASEAN Food Security Reserve. In: Chisolm, A., and R. Tyers, ed. Food security: theory, policy, and perspectives from Asia and the Pacific. DC Heath and Co., Lexington, MA.
15 At the time South Korea had yet to ratify the APTERR Agreement and was invited as an observer.
16. The flexibility accorded to ownership is intentional: Singapore and Brunei deploy private stocks (under government regulation) as their food security reserve, part of which is earmarked to APTERR. In principle this creates an opening for private sector participation in the regional reserve.

Source: Adapted from **Briones, R.M.** 2014. *Public stockholding in Southeast Asia: Review and prospects.* Philippine Institute for Development Studies. Page 10-13.
http://www.fao.org/fileadmin/templates/est/meetings/stocks/Briones_public_stockholding_in_southeast_asia.pdf

6. Assessment Test for Tool # 10: Public Foodgrain Reserves for Price Stabilization

Indicate whether each statement below is true or false.

1. Many developing countries since the 1970s, especially after the 1974 food crisis, have followed storage or stock policies to help control unfavourable price swings and to mitigate the adverse impacts on farmers as well as on low income consumers.
2. A public food reserve policy with a guaranteed minimum support price creates risk of low incomes for farmers.
3. In general, the narrower the price stabilization band, the less expensive it is to maintain for the Government.
4. In general, government procurement to build buffer stocks is more effective in controlling the price floor, than it is in controlling the price ceiling.
5. The major advantage of maintaining food reserves for price stabilization purposes is its limited costs (procurement, transportation and distribution).
6. Maintaining large food reserves for price stabilization has the potential to crowd out much needed public investment in agricultural productivity and rural infrastructure, and the possibility of discouraging private investment in the grain sector.

Answers

1. True
2. False. Food reserve policy with a guaranteed minimum support price in theory can help farmers to eliminate some of the risk associated with market access and price collapse.
3. False. In general, the tighter the price stabilization band, the more expensive it is to maintain for the Government.
4. True.
5. False. The major drawback of maintaining food reserves for price stabilization purposes is its high costs (procurement, transportation and distribution).
6. True.

7. References:

1. **Abbott, P.** 2014, Jan. *Lessons from recent stocks adjustments, and their measurements.* Paper prepared for presentation at the FAO Expert meeting on Stocks, Markets and Stability. 30-31 January 2014, Rome, Italy.

2. **Agence Française de Développement (AFD).** 2014, Feb. *Public stocks of foodstuffs in Africa,* About Development, No. 04.

3. **Antonaci L., Demeke M., Vezzani A.** 2015. The Challenges of Implementing Price and Production Risk Management in Sub-Saharan Africa, Scientific Paper 9B, ULYSSES project, EU 7th Framework Programme, Project 312182 KBBE.2012.1.4-05.

4. **Briones, R.M.** 2014. *Public stockholding in Southeast Asia: Review and prospects.* Philippine Institute for Development Studies.

5. **Dawe, D. & Timmer, P.** 2012. *Why stable food prices are a good thing: Lessons from stabilizing rice prices in Asia,* Global Food Security, Vol. 1, Issue 2, pp. 127-133.

6. **FAO.** 2014, Oct. *Food stocks and price volatility,* Committee on Commodity Problems, 70th session.

7. **FAO.** 2011. *Targeting the most vulnerable: implementing emergency reserves and other food security instruments.* Chapter 25 of *Safeguarding Food Security in Volatile Global Markets*

8. **FAO.** 2010. *Management of rice reserves stocks in Asia: Analytical issued and country experience,* by Timmer, C. P. in Commodity Market Review 2009-2010, pp. 87–120.

9. **FAO.** 2010a. *Policies for the effective management of food price swings in Africa: Managing food markets through stocks during the 2007-2008 price episode, evidence and recommendations,* Policy Brief 1.

10. **FAO.** 2006. *An Introduction to market-based instruments for agricultural price risk management,* Kang, M.G., Mahajan, N. Agricultural Management, Marketing and Finance Working Doc. 12.

11. **Galtier, F.** 2014. "Which role for storage policies in managing grain price instability? Some insights from a thought experiment". *FAO Expert Meeting on Stocks, Markets and Stability.* FAO, Rome, 30-31 January 2014.

12. **IFPRI.** 2010, Sept. *Food price stabilization: Lessons for eastern and southern Africa,* by Minot, N. Lilongwe, Malawi.

13. **IFPRI.** 2009. *Implementing physical and virtual food reserves to protect the poor and prevent market failure,* by von Braun, J. & Torero, M. Policy briefs 10.

14. **Jayne, T.S.** 2012. *Managing food price instability in East and Southern Africa,* Global Food Security, Vol. 1, Issue 2, pp. 143-149.

15. **Timmer, C. P.** 2014. *What are grain reserves worth? A generalized political economy framework.* Prepared for a workshop on the role of food reserves in coping with price volatility, held at FAO Headquarters in Rome, Italy, January 30-31, 2014.

16. **World Bank.** 2012. *Using Public Food Grain Stocks to Enhance Food Security*. Report Number 71280-GLB. Washington DC.

17. **Wright, B., & Cafiero, C.** 2011, Feb. *Grain reserves and food security in the Middle East and North Africa.* Food Security, 3(1), pp. 61-76.

Tool # 11: Disaster Assistance Programs

1. What is it? – Concepts and Characteristics

The objective of this tool is to provide a basic understanding of the nature of the disaster assistance programs and policies, how they are intended to work, their advantages and disadvantages and their relevance for risk management in agriculture.

The United Nations Office for Disaster Risk Reduction (UNISDR) defines **disaster** as "a serious disruption of the functioning of a community or a society involving widespread human, material, economic or environmental losses and impacts, which exceeds the ability of the affected community or society to cope using its own resources." Disasters happen as a combined result of exposure to a hazard, the vulnerability of those exposed and an insufficient capacity of the system to cope. Disasters, natural or human-induced, sudden or slow acting, can affect the performance of the agriculture sector as well as the livelihood of people dependent on it both directly or indirectly.

According to a FAO study (2015), during the 2003 to 2013 period, natural disasters caused USD 1.5 trillion[12] in economic damages worldwide, with more than a third of that in developing countries and about USD 80 billion in lost crop and livestock production. On average, each disaster caused a 2.6 percent loss in the national agricultural value-added in the affected countries. These losses undermine national economic growth, the performance of the agriculture sector and sustainable development.

Types of disasters – Disasters are generally grouped based on their primary cause; as natural or human-induced disasters. They can be sub-grouped depending on their nature, duration to evolve and the type of mitigating efforts required (FAO 2008) as follows.

- Natural disasters –
    - Sudden onset natural disasters – floods, earthquake, frost, locust or other quickly moving pest or disease outbreaks;
    - Slow onset natural disasters – drought, extreme weather events, El Niño, La Niña, crop or livestock damages from pest/disease infestation, etc.
- Human-induced Disasters –
    - Human conflicts such as wars, mass disturbances, etc. and
    - Socio-economic disasters such as financial/market/economic crisis, climate change, environmental contamination, etc.

Types of disaster assistance – Countries, often faced with such disasters generally have **disaster risk management (DRM)** and/or **disaster risk reduction (DRR)**[13] programs to reduce the risk, help mitigate its adverse impact on the livelihood of people and build their resilience. Such assistance programs have a strong possibility of altering the magnitude of risk in agriculture and risk behaviour of the primary producers. Therefore, the objective of this tool is to outline the various types of

---

[12] Centre for Research on the Epidemiology of Disasters (EM-DAT CRED) database.

[13] **Disaster Risk Reduction (DRR)** "refers to the conceptual framework of elements considered with the possibilities to minimize vulnerabilities and disaster risks throughout a society, to avoid (prevention) or to limit (mitigation and preparedness) the adverse impacts of hazards, within the broad context of sustainable development." (FAO 2008). While, the **Disaster Risk Management (DRM)** goes beyond DRR; it is "the systematic process of using administrative directives, organizations, and operational skills and capacities to implement strategies, policies and improved coping capacities in order to lessen the adverse impacts of hazards and the possibility of disaster. (UNISDR 2009 http://www.unisdr.org/we/inform/terminology).

disaster assistance programs and their operational goals to help build resilience, and explain how they can be useful as a tool for ARM.

There is no specific classification system but for our understanding purposes, disaster assistance programs can be grouped on the basis of their timing in relation to the disaster event and the primary purpose of the intervention.

1. **DRM Assistance Programs *(ex-post)*:**
   This includes interventions/programs/policies undertaken during and after the disaster. Although these could have some element of preparedness through pre-planning, much of the specific assistance can be ad-hoc depending on the nature of the disaster and the needs of the affected population. The primary focus of these programs is mitigation and adaptation/coping.

   a. **Mitigation** – mostly through response, recovery and reconstruction; preparedness, contingency planning, providing emergency services (including rescues), provision of basic necessities (including food, water, and medicine), rehabilitation needs, inputs for farming, shelter, repairing damages to irrigation and other farm structures,

      A major component of preparedness instruments is a strong and effective **early warning system**. The primary function of the early warning systems is to provide credible and timely warnings about food and livelihood crises to policy makers so that the necessary actions can be taken to mitigate the impact on the population. At the global level there are several prominent systems such as the FAO's Global Information and Early Warning System (GIEWS) on Food and Agriculture, WFP's Early Warning Network, and the USAID's Famine Early Warning System Network (FEWS NET). Many countries have their national and regional early warning system.

   b. **Adaptation/coping** – finding appropriate coping mechanisms, insurance or other strategies, building houses outside of flood-prone zones, adopting climate smart agricultural practices, etc.

2. **DRR Assistance Programs *(ex-ante)*:**
   This includes interventions/programs/policies undertaken generally before the disaster strikes. These can be pre-planned with a primary focus on prevention and the reduction of the disasters and/or risks of damages emanating from them before the occurrence.

   a. **Prevention** – planning capacity and public investment programs to prevent the recurrence of disasters. For example: building dams, dykes, and other structures to control the flow of the river water

   b. **Reduction** – planning capacity, institutional infrastructure - reducing deforestation, detrimental hill side cultivation, maintaining favourable grazing animals to land ratios, enforcing environmentally safe practices, etc.

Post disaster management (DRM) is crucial in saving lives and fixing damages to property and productive assets but limiting and preventing hazards (DRR) is equally as important.

The strategic goals of disaster risk reduction and management [14] are:

- Understanding disaster risk and monitoring, assessing and sharing information;
- Strengthening disaster risk governance, the coordination across relevant institutions and sectors, and the participation of relevant stakeholders.
- The development and strengthening of institutions, mechanisms and capacities to build a resilience to hazards.
- The integration of disaster risk reduction into sustainable development policies and planning.
- Enhancing multi-hazard early warning systems, preparedness, response, recovery, rehabilitation and reconstruction.

Evidently, disaster assistance is closely linked to ARM. The impact of the *ex-ante* DRR programs aimed at prevention and reduction are most likely positive in terms of the farmers' ability to manage risk, especially when combined with several other ARM tools such as climate smart agriculture, agricultural insurance, and other market-based solutions. At least in the long run, such public assistance will help reduce the production losses and increase a farmer's confidence in making investments in production, enhancing inputs and technologies.

Typically, much of the public budget is spent on the *ex-post* disaster and *ad-hoc* DRM programs which may have a complicated impact on agricultural risk management for farmers. Programs aimed at providing emergency relief and rehabilitation aid will help the farmer to safeguard his family's livelihood through most types of disasters (for example, drought, pest/diseases, etc.) in an emergency situation. However, it could also have a more perverse impact similar to "moral hazard" in the case of insurance. For example, in the case of frequent flood disasters, some farmers may continue to farm or rebuild houses in flood zones knowing that they will be bailed out in case of a disaster. From the public policy makers' point of view, this presents a challenge of providing a certain type of disaster assistance in a manner that the solution itself does not become a problem.

The source of the funding for disaster assistance can be national or international (development/financing agencies, donors, etc.). Given that, in most countries, much of the disaster management functions are managed and coordinated by the government, the 'public' assistance to deal with disasters in this study includes international or national large scale interventions in addition to prominent government programs. World Bank disaster assistance, for example, can be considered here as public assistance as it usually passes through the government. The focus is on the impact that this assistance has on farmers and the risks they face.

2. Suitability of the tool

This tool is suitable for policy makers both at a national and international level. Cost effective DRR and DRM assistance programs are especially important to countries that experience frequent natural disasters and suffer severe damages to farming and the food and agriculture sector.

This tool is particularly suitable for the agriculture sector. A recent report by FAO (2015) recommended that governments must design measures to counteract the impact of disasters specific to the crop, livestock, fisheries and forestry subsectors. Thus, a focus on disaster assistance would significantly improve farmers' abilities to manage risks in agriculture. Hence, this tool is highly suitable for agricultural risk management especially for small and marginal farmers to recover from effects of disasters and regenerate their agriculture-based livelihood.

---

[14] As outlined in the *Hyogo Framework* for Action 2005-2015 (UNISDR 2007). And the *Sendai Framework* for disaster risk reduction for 2015-2030 (UNISDR 2015).

3. Advantages

The major benefits of effective disaster assistance to agriculture is that it contributes to reduce the risk of production due to disasters and helps increase resilience to cope with disaster risk in the short to medium run. However, the formulation and implementation of appropriate country or community specific DRR and DRM policies and assistance programs is critical to bring benefits to those exposed to disaster risks. Its combination with more holistic risk management strategies is crucial for preparedness to reduce the scope of disasters.

Disasters are closely correlated with rises in food imports and drops in food exports (FAO 2015). Therefore, any reduction in the risk to the agriculture sector would bring community/national benefits in avoiding or minimizing these potential food security impacts.

According to the UNISDR (2015), international efforts, such as the adoption of the Hyogo Framework for Action, 2005-2015, have provided benefits in several ways (listed below) that are likely to reduce risk in agriculture.

- Reducing disaster risk at local, national, regional and global levels decreases mortality.
- Reducing disaster risk is a cost-effective investment in preventing future losses.
- Effective disaster risk management contributes to sustainable development, and
- Well-coordinated national and international mechanisms create necessary public awareness, political commitment and actions.

4. Disadvantages

A substantial review of the literature shows that there are recurring themes which can be considered as the limitations or lessons to be learned for disaster assistance strategies:

- Many countries spend much of their disaster assistance budget dealing with the aftermath of emergencies created by disasters and may not have an adequate amount allocated to prevention/reduction. Thus, DRR may not be well integrated into the national priority framework and not integrated into development activities.
- Not enough contingency planning and preparedness, despite a long experience of such recurrences, in some cases.
- The focus on disasters that are beyond the capacity of farmers to cope may jeopardize the need to take pro-active risk management strategies at farm community and national levels.
- Programs may exist but they may be ineffective due to weak ,national institutions and the lack of coordination at local, national, regional, and international levels.
- Inadequate funding and institutional structures - Financial reserves and emergency funding mechanisms are not effective to support preparedness, response and early recovery as required.
- International funds are usually too late, especially for emergencies. Also, donors may react and pledge after the disaster but are not willing to help in supporting *ex-ante* programs like weather index insurance.
- Effective disaster risk reduction requires community participation which may be given inadequate emphasis.
- The recovery, rehabilitation and reconstruction of the agriculture sector may not be given enough importance. For example, a World Bank (2005) study after the 2000/2001 Mozambique floods concluded that, "Although asset loss is recognized as one of the major problems faced by households post-emergency, there is still reluctance among agencies to replace capital items for individual households."

5. Examples/Case studies

EXAMPLE: Recurring Flood Disasters in Mozambique

Today, coordination is palpable and visible in Mozambique during an emergency, but not so visible in the recovery/reconstruction phase. Without a dedicated recovery program, subsequent natural disasters may further aggravate the risk to populations and infrastructure. This makes it urgent to carry out reconstruction plans more systematically. The recommendations provided by government and international stakeholders to this case study show that Mozambique is ready to further institutionalize post-disaster recovery measures.

The transition between emergency and development is a dynamic and distinct process. Although the normal planning and development process can accommodate some exceptions and emergencies, the situation must be publicly defined and articulated in order to qualify for special treatment. The transition from emergency to development should be treated as an interim phase, called Recovery. INGC is well-placed by its institutional mandate to coordinate recovery policy and programs – working with sector ministries, provincial and district governments, international partners and civil society organizations. The stakeholders in Mozambique are urged to take advantage of recently approved Disaster Management Law to debate, define and formalize the roles and responsibilities of individuals, organizations, and institutions, to ensure that recovery needs caused by recurrent disasters be fully funded, implemented and monitored.

Source: World Bank. 2014, Aug. *Mozambique: Recovery from recurrent floods 2000-2013,* Recovery Framework Case Study. Pages 1-2, 28.
https://www.gfdrr.org/sites/default/files/publication/Mozambique_August_2014.pdf

CASE STUDY 11: Caribbean catastrophe risk insurance facility (World Bank)

This case study describes how emergency pooled reserves of the Caribbean countries, assisted by donors, serve as a common risk insurance facility to help the affected countries to deal with post-disaster mitigation. The primary stakeholders of the case are the governments of these countries with the ultimate beneficiaries being the people who may be potentially affected by the catastrophe, such as hurricanes and earthquakes. Other stakeholders include the Bank and other financial service providers such as the reinsurers. The facility was established in 2007 and based on the pre-set parametric triggers it has made disbursements to several countries since. This tool is an example of a self-help disaster assistance programs at a regional level. It serves to help manage risk in agriculture given that catastrophes affect the producers engaged in the sector.

CARIBBEAN CATASTROPHE RISK INSURANCE FACILITY

The Caribbean Catastrophe Risk Insurance Facility (CCRIF) was established in 2007 basically to solve the short term liquidity problems of Caribbean governments in the aftermath of disasters. It is an exempted company under Cayman Islands laws, holding an insurance license, and is governed by a trust deed. CCRIF is a joint reserve mechanism that provides participating governments with coverage in case of disasters.

Primarily an instrument to pool resources in order to buy parametric insurance, CCRIF also covers risk from its own reserves. It is being discussed here as an example of risk pooling. To understand CCRIF, one could consider a system through which several countries agree to combine their emergency reserve funds into a common pool. If each individual country were to build up its own reserves to sustain a catastrophic event, the sum of these country-specific reserves would be much larger than the actual needs of the pooled countries in a given year. Considering that, on average, only one to three Caribbean countries is/are affected by a hurricane or an earthquake in any given year, a pool holding only the reserves for three potential payouts should be sufficient for for all pool participants. Each year, as the pool is depleted, participating countries would replenish it in proportion to their probable use. The CCRIF works in a similar manner by combining the benefits of pooled reserves from participating countries with the financial capacity of the international financial markets. It retains some of the risks transferred by the participating countries through its own reserves and transfers some of the risks to reinsurance markets where this is cost-effective. This structure results in a particularly efficient risk-financing instrument that provides participating countries with insurance policies at approximately half the price they would obtain if they approached the reinsurance industry on their own (structure of CCRIF and reduction in premium with increase in number of participating countries shown in illustrations below).

Commercial insurance is available in the Caribbean region, yet the total premium that businesses paid averaged about 1.5 percent of GDP between 1970 and 1999 while losses (insured and uninsured) amounted to only about 0.5 percent of GDP. This justifies the purchase of parametric insurance jointly, where the cost comes down substantially.

Donors provided US$67 million in start-up capital, and 16 member governments paid in US$22 million. Governments purchased parametric insurance, paying CCRIF about US$20 million in premiums for parametric insurance coverage roughly totaling US$450 million. The Facility retains responsibility for the first US$20 million of payout (backed by its capital) and

Advantages of Risk Pooling

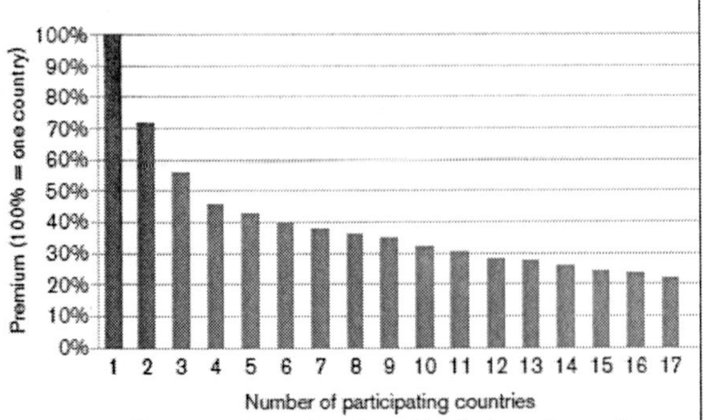

The greater the number of countries the lower the premium

Source: Esquivel[8]

Structure of the CCRIF

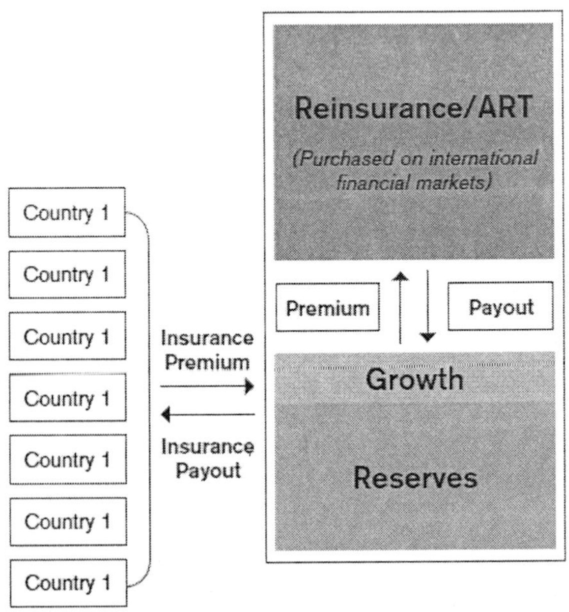

transfers the remaining exposure through reinsurance and catastrophe swaps that the World Bank intermediates. Donors expect the Facility's capital and reserves to grow and, thus, to be self-sustaining.

Within two weeks of the November 2007 earthquake, the most severe in the eastern Caribbean in 30 years, the Facility paid about US$1 million to St. Lucia and Dominica . It paid US$6.3 million to the Turks and Caicos Islands after Hurricane Ike in September 2008.[1,2,5] There have also been disasters that did not trigger the parameters set: Hurricane Dean in 2007 caused considerable damage in Jamaica because of rain, but no payout occurred because wind speed was the parametric trigger. Similarly, the cumulative effect of the 2008 hurricanes in Haiti was devastating, but the winds were not strong enough to trigger a payout. The 7.0 magnitude earthquake that struck Haiti on January 12, 2010, was of sufficient magnitude to trigger the full policy limit for Haiti's earthquake coverage purchased under the Facility. Based on calculations from the preliminary earthquake location and magnitude data, Haiti will receive US$7.8 million, the maximum payout under its earthquake policy. This is about 20 times its premium for earthquake coverage of US$385,500. Although shaking was felt in Jamaica, another CCRIF-covered country, it was insufficient to generate any loss under the parametric index.

Participating countries can decide on coverage as follows:

- Perils Covered: Hurricane (protection against hurricane wind damage) and earthquake.
- Coverage Limit: The maximum possible payout for all claims in any one policy period, not to exceed US$100 million and specified for each insured peril.
- Attachment Point: Level of government loss that triggers the policy, being the equivalent of a deductible. It can also be expressed in terms of the return period of experiencing the level of loss or more. The Attachment Point cannot be set at loss levels with return periods of less than 15 years. (For CCRIF insurance purposes, the only valid method to calculate the amount of government loss is the parametric loss model specified in the policy itself.)

Strengths

CCRIF allows the participating countries to pool their risks and resources, creating a more diversified portfolio. While this increases the funds at the disposal of affected country as a retention mechanism, it also reduces the cost of insurance. In the case of CCRIF, the premium is reduced by half as compared to what a country would pay for individual insurance.

CCRIF is backed by donor funds held by the World Bank in a multi-donor trust fund. These additional resources help reduce the CCRIF's dependence on the reinsurance market by increasing its own reserves.

Scope for Improvement

A review of parametric trigger mechanisms is needed. Non-payment of any damages in Hurricane Dean in Jamaica has shown that only wind speed measurements as triggering devices are not sufficient and more frequent events need to be covered giving wider coverage.

The procedures for a multi-country fund need to be streamlined. Another important requirement in a multi-country fund is to have very clear and transparent procedures with high visibility and communication at all levels. This precludes any confusion in the aftermath of disasters as to whether a country could have received payments.

CONCLUSION

The choice of instruments in disaster risk financing depends on many factors. One way of looking at it is to classify the disasters in terms of their expected severity and frequency. More frequent disasters with low expected severity (e.g., recurring floods) are better financed by retaining the risk, as the cost of transferring such risk will be disproportionately high compared to the expected damages or payments. On the other hand, risk associated with low frequency–high severity disasters (e.g., major earthquakes) is best transferred to the international reinsurance market, as government may not have the capacity and resources to sustain the damages caused by such disasters.

Another way of looking at the issue is to map the risk financing tools by social or geographical relevance. Tools for the individual or household level naturally will differ from those for national governments.

Taking in to consideration the scope of the topic, this note focused only on risk retention and financing tools relevant for national governments or international bodies. The mode chosen was to study successful examples in the field and to assess their strengths and weaknesses. To the extent that this review is fungible, the countries will benefit from mixing and matching the models and instruments to their specific geo-political context. Finally, as the case studies in this note indicate, the uptake of risk retention and risk transfer mechanisms has made countries more resilient to natural disaster. ■

REFERENCES

1. Dumitru, D. 2010. Mitigating the Adverse Impacts of Natural Disasters on the Philippines Economy: A Study of Disaster Risk Financing Options. Mimeo.
2. World Bank. Catastrophe Risk Financing in LCR: Recent Solutions and Challenges Ahead. IBRD Results Working Paper. Washington, DC.
3. World Bank. FONDEN: National Disaster Fund in Mexico. Note Prepared by World Bank Team for Indonesia Mission. Washington, DC.
4. Interview with Rubem Hofliger Topete, Director General, FONDEN.
5. Ghesquiere, F. and F. Ramirez Cortes. 2008. Mechanisms for the Mobilization and Allocation of Financial Resources in Post-Disaster Solutions in Costa Rica. Washington, DC: World Bank.
6. Sanghi, A. et al. 2010. Unnatural Disasters, Natural Hazards: Effective Prevention through an Economic Lens. Washington, DC: World Bank.
7. Background Note on CAT DDO available at World Bank website

   http://web.worldbank.org/WBSITE/EXTERNAL/NEWS/0,,contentMDK:21670477~pagePK:64257043~piPK:437376~theSitePK:4607,00.html
8. Esquivel, M. Caribbean Catastrophe Risk Insurance Facility (CCRIF): A Joint Reserve Mechanism for Island States in the Caribbean. Washington, DC: World Bank.
9. World Bank. 2008. A Review of CCRIF's Operation after Its First Season. Washington, DC.
10. World Bank. 2008. The Caribbean Catastrophe Risk Insurance Facility: Providing Immediate Funding after Disasters. Washington, DC.
11. Cummins, J.D. and O. Mahul. 2008. Catastrophe Risk Financing in Developing Countries: Principles for Public Intervention. Washington, DC: World Bank.
12. O'Donnell, I. 2009. Practice Review on Innovations in Finance for Disaster Risk Management. Geneva: ProVention Consortium.
13. Climate and Disaster Risk Solutions-Equipping Africa with Technology and Tools to Manage Natural Disaster Risk, Rockfeller Foundation and UNWFP.

Source: Adapted from **World Bank**. 2010. Disaster Risk Financing: Case Studies. Working Paper Series No. 23. Disaster Risk Financing: Case Studies By Sandeep Poundrik. Pages 10-12. http://www-wds.worldbank.org/external/default/WDSContentServer/WDSP/IB/2011/03/28/000356161_20110328042850/Rendered/PDF/604560BRI0231R10BOX358322B01PUBLIC1.pdf

6. Assessment Test for Tool # 11: Disaster Assistance

As a policy maker, you are required to develop and implement effective methods to deal with disasters. Your specific area is known for its frequent destructive floods. There has been a recent flood and action is required. List the following actions in order of urgency in the above given situation and also indicate whether this action would come under the disaster risk reduction (DRR) strategy or disaster risk management (DRM) strategy.

Action/Activity:

a. Getting the entire community involved in order to quicken the recovery stage.
b. Planning and constructing dams to prevent the recurrence of another flood.
c. Provide emergency resources and services such as drinking water, food and medicine.
d. Provide temporary shelter for those affected by the flood.
e. Commence building houses outside of flood prone zones.

Answers

Priority and the nature of strategy

First – Action c (DRM)
Second – Action d (DRM)
Third - Action a (DRM)
Fourth – Action e (DRR)
Fifth – Action b (DRR)

7. References:

1. **FAO.** 2015. *The impact of disasters on agriculture and food security.* A call for action to build resilient livelihoods.

2. **FAO.** 2013. *Mainstreaming disaster risk reduction into agriculture: A case study from Bicol Region, Philippines,* by Mascariñas, A. M.

3. **FAO.** 2008. The State of Food Security in the World 2008. Rome.

4. **FAO.** 2008a. *Disaster risk management systems analysis: A guide book,* by Baas, S., Ramaswamy, S., Dey de Pryck, J., Battista, F.

5. **Little, P. D., Smith, K., Cellarius, B. A., Coppock, D. L. and Barrett, C.** 2001, *Avoiding Disaster: Diversification and Risk Management among East African Herders.* Development and Change, 32: 401–433.

6. **Turnbull, Marilise; Sterrett, Charlotte and Hilleboe, Amy.** 2013. Toward Resilience - A Guide to Disaster Risk Reduction and Climate Change Adaptation. Practical Action Publishing Ltd, UK, Catholic Relief Services – United States Conference of Catholic Bishops.

7. **UNISDR** (United Nations Office for Disaster Risk Reduction). 2015. *Sendai framework for disaster risk reduction 2015-2030.* 43291_sendaiframeworkfordrren

8. **UNISDR.** 2014, May. *Fifth Africa regional platform for disaster risk reduction.*

9. **UNISDR.** 2010. *Local Governments and Disaster Risk Reduction: Good Practices and Lessons Learned,* United Nations Secretariat to the International Strategy for Disaster Reduction, Geneva.

10. **UNISDR**, "2009 UNISDR Terminology on Disaster Risk Reduction", Geneva, May 2009

11. **UNISDR.** 2008. *Disaster preparedness for effective response: guidance and indicator package implementing priority five of the Hyogo framework.*

12. **UNISDR.** 2007. *Hygo framework for action 2005-2015: Building the resilience of nations and communities to disasters,* Extract from the final report of the World Conference on Disaster Reduction.

13. **USAID.** 2010, Jun. *Disaster risk reduction (DRR) programs. Fact Sheet No. 1, Fiscal Year 2009.*

14. **World Bank.** 2014, Aug. *Mozambique: Recovery from recurrent floods 2000-2013,* Recovery Framework Case Study.

15. **World Bank.** 2006. *Hazards of nature, risks to development: An IEG evaluation of World Bank assistance for natural disasters.*

16. **World Bank.** 2005, Apr. *Learning lessons from disaster recovery: The case of Mozambique,* Wiles, P., Selvester, K., Fidalgo, L. Working Paper Series, No. 12.

17. **World Bank.** 2010. Disaster Risk Financing: Case Studies. Working Paper Series No. 23. Disaster Risk Financing: Case Studies By Sandeep Poundrik. Pages 10-12

Tool # 12: Social Protection and Productive Safety Nets

1. What is it? – Concepts and Characteristics

The objective of this tool is to provide a basic understanding of the nature of social protection and productive safety net programs and policies, how they are intended to work, their advantages and disadvantages and their relevance for risk management in agriculture.

Social protection refers to a set of policies and programs aimed at supporting poor and vulnerable people by promoting efficient labour markets, diminishing people's exposure to risks, and enhancing their capacity to manage economic and social risks such as unemployment, exclusion, sickness, disability and old age. Historically, in many developed and developing countries, social protection or social safety nets[15] have played an important role in protecting the livelihoods of weaker sections of the population but has gained more significance in developing countries in the 21st century.

With the successful adoption of some large-scale programmes in countries like Brazil, Ethiopia, India and Mexico, social protection has gained a new focus in combating poverty, hunger and inequality (FAO 2015a). The use of social safety nets has been growing and as of 2015, every country in the world has had at least one social safety net program in place. According to the World Bank survey (2015) in Africa, 40 countries (out of 48 in the region) have unconditional cash transfers, doubling since 2010.

Given the close links between social protection and risk in agriculture, this tool is prepared to provide a basic understanding of the nature and potential contributions of social protection and productive safety net programs to ARM. This tool is particularly targeted at the public policy and decision makers regarding the welfare of those engaged in agriculture and exposed to risks in their production, on top of their own problems and vulnerability due to poverty and inadequate resources.

Social protection is offered in the following three general categories[16]:

- **Social assistance** - The most common programmes are:
    1. Unconditional transfers, i.e. programmes that distribute **cash or vouchers**, or are **in-kind (such as food)**, food without anything required of the recipient;
    2. Conditional transfers, which may otherwise be identical to unconditional transfers except in that they require recipients to meet some specified conditions, typically to improve the human resources of their children;
    3. Public works programmes, also referred to as **cash- or food-for-work, or guaranteed employment programmes**, which require beneficiaries to work to create or maintain household or community assets.
- **Social insurance** is provided through transfers by the government, non-profit agencies, and the community or family members. These programs include health insurance or unemployment insurance and help to manage risks associated with unemployment, ill health, disability, work-related injuries and old age.
- **Labour market interventions** are policies and programs designed to promote employment, the efficient operation of labour markets and the protection of workers. Social programmes

---

15 Although the terms social protection and social safety net are used here interchangeably, safety nets are part of a more holistic approach of social protection system. For further discussion, see Devereux, 2009; World Bank 2015.

16 Concepts are based on FAO 2015a, Slater *et al.*, 2010.

in this category provide unemployment benefits, build skills and enhance workers' productivity and employability.

Some social support may be aimed at creating development for vulnerable producers to safeguard their production and livelihoods. Such support, called a **productive safety net program** (PSNP), is practiced in some developing countries (e.g. Ethiopia). The most common elements of these programs are:

- **Social development** (school feeding, nutrition programmes, education stipends, fee waivers, etc.),
- **General economic development** (public works programmes, energy/fuel subsidies, etc.) and,
- **Sector specific economic development:** assistance such as PSNP for farmers (farm input subsidies, conservation and other investment grants, free or subsidized veterinary services/vaccinations, risk insurance and financing etc.).

Social protection programs aimed at reducing poverty and vulnerability in agriculture, in general and agricultural PSNPs in particular, help farmers to reduce risk and provide ways to manage risk better in agriculture. These programs also help to cope with the risk of farm income variability by smoothening consumption and preventing the forced adoption of perverse/destructive coping strategies. Social protection in relation to agriculture, food security and nutrition helps build resilience, i.e. the inherent capacities of individuals, groups, communities and institutions to withstand, cope, recover, adapt and transform in the face of specific shocks, in agriculture (FAO 2015a).

2. Suitability of the tool

This tool is particularly suitable and highly effective for smallholder marginal farmers when it is combined with the farmer's own private risk management tools such as agricultural insurance, crop/enterprise diversification and other modern market-based approaches to risk management in agriculture.

Historically, farm input support programs aimed at smallholder farmers have been implemented by many countries but were later abandoned for their potential market-distorting effects. However, following the frequent episodes of soaring global food prices and serious food insecurity occurrences in recent years, a variety of input support, particularly in the form of fertilizer and seed subsidies, has become popular in many developing countries (Demeke *et al.* 2014).

A recent FAO publication (2015a) cites specific examples from Ethiopia, Lesotho, Rwanda and other countries and shows that the social protection and agricultural support interventions can be more effective together than in stand-alone programs. The linking of several programs can also help improve targeting and adjusting various interventions and improve the suitability of the tool.

As far as the suitability of the specific programs is concerned, cash transfers appear to be superior in terms of the efficient achievement of objectives as compared to in-kind transfers, therefore, cash-based safety nets seem to have become more appealing, even in developed countries (World Bank 2012).

3. Advantages

There are a number of actual and potential synergies and conflicts between smallholder agricultural policies and social protection policies in Africa (Devereux 2009).

Social protection has obvious benefits to vulnerable families when they receive support for a given period which can allow them to gain their own resilience. The existence of such social safety net programs may provide added security and allow farmers to take risks in making necessary investments in productivity enhancing inputs and technology.

Therefore, when social protection and farm production supports are combined effectively, they can lift the targeted farmers out of the food insecurity and poverty trap. Fertilizer subsidies can raise input use, resulting in an increase in food production within a relatively short amount of time. The promotion of fertilizer usage along with appropriate public development assistance for research and development, extension services, irrigation, marketing and credit facilities and other rural infrastructure has been proven to generate sustained and efficient gains in agriculture. Among the various cases, Asian countries such as Bangladesh, India, Indonesia and Pakistan are seen as success stories (FAO 2015a).

Farm input support programs, when properly targeted at low-income, small family farmers can help to increase their farm income and reduce its variability while achieving the dual goal of social equity.

4. Disadvantages

There are several limitations or issues of concern with respect to the nature and size of social safety net programs and their implementation (FAO 2015a, 2015b). Some of the major ones are listed below:

- Certain forms of social protection, for example cash transfers, can have an inflationary impact on local markets; specific local case examples such as Ethiopia, Kenya, Uganda, Zambia, are found in the literature review.
- In some cases, food aid distributions as a social protection measure can have a dampening effect on local markets and act as a disincentive to farmers.
- The targeting of individuals/families for social support programs can be difficult. Several examples from Zambia and Malawi are cited (FAO 2015a). Thus, the benefits of such programs may not always be towards the intended individuals. Empirical studies have found that farm subsidies may benefit large scale farmers proportionately more than the small holder farmers.
- Timely adjustment of the target population after the occurrence of a hazard has proved to be challenging. For risk management purposes social protection targets need to go beyond situations of chronic poverty or food insecurity to truly transitory social protection after shocks.
- The effectiveness of input subsidies can also be questionable, especially when measured in terms of the opportunity cost of limited government budgets. The fiscal sustainability of input subsidies is a major issue for many developing countries.
- The impact on risk behaviour of farmers may not be guaranteed especially when the farmers do not believe the certainty of support programs.
- While designing social safety net programs, policy makers have to keep in mind that large scale public support programs do not crowd out private sector involvement and that the

development of the market and financial sectors (including insurance products markets) is not adversely affected.

5. Examples/Case studies

EXAMPLE: Ethiopia's "Productive Safety Net Programme" (PSNP)

The PSNP, launched in 2005, aims to provide transfers to food insecure households, strengthening livelihoods and shifting away from emergency food aid distribution. The programme promotes "graduation", defined as the moment when households can meet basic food needs and are able to face modest shocks. The number of people covered by the programme has increased from 5 to 8 million in 2008.

The programme provides transfers (cash and/or food according to regions) in return for work, and provides direct support to households with no labour or other means of support. Transfers were un-indexed, but in response to rising food prices and declining value of cash transfers in 2008, the payment was raised from Birr 6 to Birr 8 per day (FAO 2009; Sabates-Wheeler and Devereux 2010).

Source: Adapted from FAO. 2011. *Social safety nets and the food security crisis. Box 2*
http://www.fao.org/fileadmin/user_upload/ISFP/Social_safety_nets.pdf

CASE STUDY 12: India's Mahatma Gandhi National Rural Employment Guarantee (FAO)

This case study provides an example of the provision of guaranteed rural employment to those who need an income to maintain their livelihood in India. Millions of people participate in the scheme each year. It has been one of the longest running large scale social protection programs in the world and it has helped people to combat rural poverty and hunger. The case study explains how it works. It helps build the resilience of those with a shortfall in their own farm production or labour earnings from their usual employment and thus acts as a risk management strategy for smallholder and marginal farmers.

BOX 7

India's Mahatma Ghandhi National Rural Employment Guarantee Act

The MGNREGA is a historic piece of legislation aiming at two interlinked goals. Its rights-based approach views employment as a right of the citizen to be delivered by the state. The first goal is to ensure livelihood security to rural residents by providing at least a hundred days of guaranteed wage employment in a fiscal year to every household with an adult member willing to do unskilled manual work for a minimum wage. The second goal is to mobilize existing surplus labour in the countryside, unleash productive forces and generate more economic growth in rural areas.

The Act came into force on 2 February 2006 and was implemented in phases to cover all rural districts within three years. At its peak in 2010/11, it covered more than 55 million rural households, about a third of all rural households, generating 2.6 billion days of employment in that year. It is a relatively inexpensive programme: even at its peak, total spending came to less than 1.0 percent of GDP, and it currently accounts for less than 0.5 percent of GDP. The programme is large, but has been unevenly implemented across states. The peak years of MGNREGA performance, in terms of both financial and employment indicators, were 2009/10 and 2010/11; there has been a general decline in most states since then (Ghosh, 2014).

The programme is designed to work as follows (MORD, 2013). Adult members of all rural households willing to do unskilled manual work register with the local *Gram Panchayat* (the lowest-level elected body). Each household is entitled to 100 days of employment per year, although the programme has only provided 40–50 days of employment per household each year so far. The registered household is issued a job card. A written application for work is the basis for a guarantee of providing employment within 15 days. If this is not provided, the state is supposed to pay an unemployment allowance (of half the wage rate) to the beneficiary. At least a third of the beneficiaries of the scheme must be women. In practice, women have accounted for around half of total beneficiaries. Facilities – such as crèches for children, drinking water and shade for rest – are supposed to be provided. The cost of projects, excluding wages for beneficiaries, cannot be more than 40 percent of total costs. Contractors and use of labour-displacing machinery are prohibited.

Source: Ghosh, 2014.

Source: FAO 2015a. The State of Food and Agriculture Social protection and agriculture: breaking the cycle of rural poverty. Page 28. Rome. http://www.fao.org/3/a-i4910e.pdf

6. Assessment Test for Tool # 12: Social Protection and Safety Nets

Fill in the blanks in the following statements using the words given in the list at the bottom.

1. __________ aims at supporting poor and vulnerable people by enhancing their capacity to manage economic and social risks.
2. The existence of social safety net programs may provide added security and allow farmers to take risks in making necessary investments in __________________inputs and ____________.
3. The productive safety net programs (PSNPs) are considered a part of social protection policies given that they aim to improve household _______________ and reduce ____________.
4. PSNPs for farmers include farm input _________, _______________, free or subsidized __________ services and ________ insurance.
5. Labour market ________ are policies or programs that promote ___________, efficient operation of labour markets and the _________ of workers.

Key missing words:

employment
food security
hunger
interventions
investment grants,
productivity enhancing
protection
risk
Social protection
subsidies,
technology
veterinary

Answers

1. *Social protection* aims at supporting poor and vulnerable people by enhancing their capacity to manage economic and social risks.
2. The existence of social safety net programs may provide added security and allow farmers to take risks in making necessary investments in *productivity enhancing* inputs and *technology.*
3. The productive safety net programs (PSNPs) are considered a part of social protection policies given that they aim to improve household *food security* and reduce *hunger*.
4. PSNPs for farmers include farm input *subsidies, investment grants,* free or subsidized *veterinary* services and *risk* insurance.
5. Labour market *interventions* are policies or programs that promote *employment* efficient operation of labour markets and the *protection* of workers.

7. References:

1. **Demeke, M., Spinelli, A., Croce, S., Pernechele, V., Stefanelli, E., Jafari, A., Pangrazio, G., Carrasco, G., Lanos, B. & Roux, C.** 2014. *Food and agriculture policy decisions: trends, emerging issues and policy alignments since the 2007/08 food security crisis.* Rome, FAO.

2. **Devereux, S.** 2009, Jan. *Social protection for agricultural growth in Africa,* FAC Working Paper No. SP06. Future Agricultures. Centre for Social Protection.

3. **FAO** 2015a. The State of Food and Agriculture Social protection and agriculture: breaking the cycle of rural poverty. Rome.

4. **FAO.** 2015b. *Strengthening coherence between agriculture and social protection: Zambia country case study report,* by Harman, L.

5. **FAO.** 2013, May. *Integrating risk management, Tools and policies into CAADP: Options and challenges,* by Antonaci, L., Demeke, M., Soumare, M.S.

6. **FAO.** 2011. *Social safety nets and the food security crisis.* http://www.fao.org/fileadmin/user_upload/ISFP/Social_safety_nets.pdf

7. **Ghosh, J.** 2014. Social protection programmes in India: An overview of recent experience with different types of schemes. Paper prepared for the Food and Agriculture Organization of the United Nations. Rome, FAO. Unpublished.

8. **Sabates Wheeler R. and Devereux S.** 2010. *"Cash Transfers and High Food Prices: Explaining Outcomes on Ethiopia's Productive Safety Net Programme"*, Working Paper 004, Future Agriculture.

9. **Slater, R., Holmes, R., Farrington, J., McCord, A. & Hagen-Zanker, J.** 2010. *Linking agriculture and social protection toolbox. A2: Concepts and framework.* London, ODI.

10. **World Bank.** 2015. The State of Social Safety Nets 2015. Washington, DC.

11. **World Bank.** 2012. Managing risk, promoting growth. Developing systems for social protection in Africa. The World Bank's Africa social protection strategy 2012–2022. Washington, DC.

PARM

PLATFORM FOR AGRICULTURAL RISK MANAGEMENT

www.p4arm.org